Cell Structure

The Authors

Peter G Toner BSc MB ChB

Department of Pathology, Western Infirmary, Glasgow.
Formerly of the Department of Anatomy, University of Glasgow.

Katharine E Carr BSc PhD

Bio-Engineering Unit, University of Strathclyde.
Formerly of the Department of Anatomy, University of Glasgow.

G M Wyburn DSc FRCP(Glasg) FRSE

Regius Professor of Anatomy, University of Glasgow

Cell Structure

An Introduction to Biological Electron Microscopy

Peter G Toner
Katharine E Carr

Foreword by G M Wyburn

Department of Anatomy University of Glasgow

E & S LIVINGSTONE Ltd, Edinburgh & London, 1968

SBN   443   00580

Printed in Great Britain

Foreword

Asked to comment on a high resolution electron micrograph of muscle the artist replied 'I don't know what it is but it is beautiful.' Anatomists share with the artist the appreciation of structure in its own right as an abstraction like a painting, a fine piece of music, or the beauty of the sea anemone.

The electron microscope has revealed to the anatomist new concepts of pattern within the size range between the chemical unit, the molecule and the biological unit, the cell. The myelinated nerve fibre, the collagen fibre, the section of a cilium, have no counterparts in structure at naked eye level.

Today it is customary for those who wish to exalt anatomy as a dynamic and modern science to proclaim its functional orientation and experimental approach. For the avant-garde biologist structure has a low rating, the frontiers are function. Yet the physicist and the astronomer are content to leave the function of the particle and the purpose of the universe to the philosopher.

Requested to design the tools for cell secretion or for the transduction of photo-chemical reaction to receptor potential, it is unlikely that anything resembling endoplasmic reticulum with Golgi apparatus or retinal cones and rods would be blueprinted. Function cannot be interpreted in terms of structure. Given the structure, the function will in time announce itself. Thus the anatomist should be well content with his task of exploration of the manifold arrangements of structure with their emergent properties that summate to homo sapiens and the revelation of the diversity of pattern which has evolved seemingly independent of any unifying functional directive.

In the early 50's it was, at least in Britain, an exciting experience to produce electron micrographs of biological material sufficiently good to tell a story and the proud day of our first such publication in 1954 has never quite been recaptured. At that time the thickness of the sections cut with American safety razor blades on a modified Spencer rotary microtome was probably to be measured in fractions of a micron rather than in terms of Ångstroms'. Yet given the fortunate section they were the correct thickness to demonstrate that nuclear pores were something more than gaps in the membrane.

The 47 micrographs produced to illustrate this book testify to the progress of the intervening fourteen years, in part due to more sophisticated apparatus but only rendered possible by technical skill and experience. The intention here is not to give a complete survey of the fine structure of all tissues and organs but rather to select and emphasise such common denominators of cellular structure as the cell membrane and its modifications, the mitochondria, the intracellular apparatus associated with specific tasks, protein and ion secretion, absorption, and the structural adaptations concerned with transmission, conduction and contraction.

v

The electron microscope has rationalised rather than complicated our concepts of tissue and cellular structure and given meaning to pristine empirical descriptions such as 'striped' muscle or 'striated' border. The educational theorist might object that the proper heuristic approach is a vertical integration of the gross, microscopic, and fine structure, but experience favours in the first instance the horizontal presentation leaving the learning process to do the integration. Thus, histology included in the textbook of anatomy tends to be subordinated and electron microscopy interpolated in the textbook of histology loses much of its impact. To arrest the attention of the student this third dimension of structure must be given equal status in the curriculum with other aspects of the subject and this is best accomplished by the assembly of suitable electron micrographs with appropriate texts as in this first British textbook of its kind primarily directed at the medical student.

In the beginning there was anatomy. Anatomy begat anthropology, physiology, pathology and biological chemistry. In time, perhaps sooner rather than later, the anatomist and the physical chemist will share a common nomenclature and enjoy a common understanding. Today anatomy has given medicine a new perspective in structure, a new base line of the normal for the better understanding of the abnormal and thus a new impetus to progress in the treatment of disease.

G. M. Wyburn

Glasgow, 1968.

Preface

This book is intended as a simple introduction to biological electron microscopy. In it we have set out to do three things : firstly, to present the fine structure of the cell and a number of the more interesting specialisations of cell structure ; secondly to provide enough technical information to satisfy the first questions of the more interested student and to indicate to him the potential uses and limitations of electron microscopy ; finally to help the beginner to approach the examination and interpretation of an unknown micrograph in a systematic way.

We have not attempted to compile a comprehensive reference work on fine structure since texts of this type are already available. Nor have we tried to present a manual of technique but instead have limited this section to give background information upon which an interested student might subsequently build. We have assumed that the study of fine structure will form part of a more general biological training and our limited treatment of functional aspects is not intended to take the place of the detailed study of biochemistry.

We have become convinced of the need for a book of this kind from our contact with students at the early stages of their medical studies and also from our experience of the needs and interests of students attending extramural classes on biological electron microscopy at Glasgow University. We believe that a working knowledge of fine structure may soon be as important to the biologist as a knowledge of histology and that a systematic introduction to the subject is best provided in the present form, rather than as supplements in a larger text of anatomy or histology. We hope that the book will prove of interest not only to the medical and biology students at whom it is primarily aimed, but also to those now past their student days who have not been exposed to any formal teaching of the elements of fine structure. We would like to feel that this book might help any, who for this reason regard fine structure with misgivings, to feel more at ease when confronted with the increasing numbers of electron micrographs appearing in the pages of the scientific press.

We are indebted to Professor G. M. Wyburn for the use of the facilities of the Department of Anatomy and for his advice and helpful criticism, not only during the preparation of this book, but on many occasions in the past. The electron micrographs with which the book is illustrated were taken by us using the Philips E.M. 200 electron microscope of the Department of Anatomy at Glasgow University. Miss Jean Hastie and Miss Pauline Semple assisted in collecting and processing the tissues and in preliminary screening. Miss Margaret Hughes gave invaluable photographic support and prepared all of the final prints. Miss Jane Young of the Department of Anatomy and Mr. D. Lang of E. & S. Livingstone produced the line drawings with skill and care. We are most grateful for the assistance provided in these different ways, without which our own work would have been immeasureably increased. We would also like to thank the staff of E. & S. Livingstone for their co-operation and assistance at all stages in the production of the book and we are most grateful to Mr. F. Dubrey of Scottish Studios for the care he has taken with the reproductions of our electron micrographs.

A number of our friends and colleagues have given us their help and criticism. We are particularly grateful to Dr. J. P. Arbuthnott and to Dr. I. A. Carr, and we would like also to thank Drs. R. B. Goudie, W. A. Harland, E. Arbuthnott, D. Graham, K. C. Calman, J. S. Dunn, A. R. Henderson, A. M. MacKay, R. F. Macadam, Mr. A. Martin, Miss J. Rentoul, and Mr. R. Young for their comments at different stages. Professor J. R. Anderson, Western Infirmary Department of Pathology, Glasgow University, has kindly given his encouragement and interest. We accept all responsibility for the remaining shortcomings in the text and for inadequacies in the micrographs, but we hope that they will not prevent the book from being of use to those with an interest in cell structure.

P. G. T.
K. E. C.

Glasgow, June 1968.

Contents

CHAPTER 1

The Study of Biological Structure

The study of structure is an essential basis of biology. Biological structure is examined at different levels, first by the direct examination of specimens with the unaided eye, then by the use of technical aids, such as the microscope, which extend the range of vision. Since structural details of different dimensions are shown by these various means it is necessary for the observer to adapt to different scales of measurement. The centimetre and the millimetre were too large to serve the purposes of microscopy and the micron (μ) became the most commonly used unit. One thousand microns are equal to one millimetre.

For many years the light microscope has been the most important of these technical aids used for detailed structural investigation in biology. Magnifications of up to two thousand times are possible by light microscopy, allowing details to be seen which are far beyond the reach of unaided vision. The cells or tissues to be examined are preserved using chemical fixatives such as formalin and supported by embedding in blocks of paraffin wax. Histological sections, thin slices of the supported tissue, can then be cut on a microtome with a metal knife and mounted on a glass slide. This thin specimen can be stained with coloured dyes and examined by transmitted light.

There is one important limitation to light microscopy. The details of biological structure which can be made out by the use of the light microscope are limited by the physical nature of the light itself, however perfect the microscope and the specimen may be. The detail which can be made out using the light microscope is limited by the wavelength of visible light which ranges from 0·4 to 0·7 μ. For this reason, two particles less than 0·2 μ apart in the specimen will not be distinguished, or resolved, as separate images, but will appear as a single blurred structure. Such details are said to be beyond the limit of resolution of the light microscope.

There are other instruments apart from the light microscope which assist the study of structure, particularly molecular structure. By the use of X-ray diffraction techniques, the three dimensional structure of complex molecules such as myoglobin and the enzyme lysozyme has been elucidated. The biochemist investigates the relationship between the structure and the functions of the molecular components of tissues while the physical chemist can define the relationships of the atoms within a single molecule. For some time, however, there was no link between the details of tissue organisation seen by the light microscope and the details of molecular structure and architecture revealed by these analytical techniques.

The electron microscope has provided this link. By using electrons instead of light to form the image of the specimen, the restrictions on resolution can be greatly reduced, since the wavelength of the electron beam in the operating conditions of the electron microscope is many times smaller than the wavelength of visible light. The difference in scale between light and high resolution electron

A

microscopy is so great that new units of measurement are needed. Decimal fractions of the micron are often used and the dimensions of small structures seen with the electron microscope may be measured in terms of thousandths of a micron, or milli-microns (mμ). Even this unit is too large at times, since the modern electron microscope can resolve two points which lie as little as 0·2 or 0·3 mμ apart. The unit most commonly used at the limits of high resolution electron microscopy is the Ångstrom unit (Å). There are 10,000 Å in one micron and 1 mμ is therefore equal to 10 Å. The practical limit of the modern electron microscope is at present between two and three Ångstroms, still far from the theoretical limit imposed by the wavelength of the electron beam. Lens design and specimen preparation are now the main limiting factors in resolution.

The permanent record of a specimen produced by the electron microscope is in the form of a photographic plate which can be enlarged to produce the electron micrograph, a print in black and white on photographic paper. The modern microscope can form an image at a magnification of 500,000 times, which can be increased to 2,000,000 times by photographic processing, although it is important to remember that the resolution of an electron micrograph is a more important measure of its value than its magnification. Since such high magnification is possible, the observer already familiar with histology at light microscopic level may find it difficult to adjust to the change in scale which is required. In practical terms the differences between light and electron microscopy are reflected in the time which they consume. To record on photographic plates the area of tissue scanned by the light microscopist in a few minutes could take weeks of work with the electron microscope. In addition, the high resolution of the electron microscope has forced a revolution in specimen preparation. New fixatives such as osmium tetroxide and glutaraldehyde have been introduced which cause the minimum of structural damage to the tissue. Wax embedding has been replaced by plastic embedding, making possible the production of much thinner sections. Ultramicrotomes have been designed which can cut sections as thin as 500 Å, using knives made from the freshly broken edge of a piece of glass.

The foundations of biology have been laid by microscopy. Despite the limited resolution of the light microscope, the theory of the cell as the unit of life came from its use during the eighteenth and nineteenth centuries by the early histologists. The description of the nucleus and the cytoplasm, the main parts of the cell, became accepted and elementary subcellular components, the mitochondria and the Golgi apparatus, were discovered, although their significance was disputed. The limited resolution of the light microscope made it impossible to resolve arguments concerning these fine details of cell structure.

With the introduction of electron microscopy, new concepts of subcellular organisation have evolved. The detailed information now available has become known as 'fine structure' or 'ultrastructure' and the investigation of fine structure is already an important part of biology. Fine structural features common to all cells have now been found by electron microscopic examination. Distinctive appearances have been described in association with specific cell functions and by combining fine structural and biochemical techniques it has become possible to determine the composition and function of the different components of the cell and to assess the contribution each makes to metabolism. In the following chapters some of the results of these recent studies of the fine structure of the cell will be described.

CHAPTER 2

Biological Membranes and the Cell Surface

THE MEMBRANE AND ITS FINE STRUCTURE
(*Plates 2, 5 and 44 illustrate the structure of the cell membrane.*)
A *membrane* can be defined as a tenuous partition or interface between two phases
of the substance of the cell, or between the cell and its environment. The demonstration by the electron microscope of the extent and specialisation of biological
membrane systems has made it clear that membranes are essential in the organisation of cell structure and function.

The membrane previously studied in the greatest detail by cytologists is the
cell membrane, or plasma membrane, which forms the outer boundary of the cell,
separating the cytoplasm from the extracellular environment. For many years it was
realised that some form of partition, too thin to be observed directly using the light
microscope, must exist at the cell boundary. The swelling of cells when placed in
hypotonic solutions indicates the presence of a semi-permeable cell membrane
which allows the passage of water into the cell while resisting the outflow of large
molecules from the cytoplasm. The resilience and elasticity of the cell membrane
can be demonstrated by micromanipulation. If the membrane is torn or punctured
the escape of cell contents can be observed. Thus there is convincing evidence of
the existence of some form of structural partition between the interior of the
cell and its surrounding environment. Cell membranes in bulk have been obtained
from red blood corpuscles, although these may not be entirely typical of cell membranes in general. Red corpuscles suspended in distilled water swell as water
passes in, driven by osmotic pressure, and finally rupture. The membranes can
then be collected for detailed physical and chemical analysis.

Electron microscopic examination has finally provided a visible demonstration
of a bounding cell membrane with a basic structure common to every form of cell.
The thickness of the cell membrane as seen in the electron microscope varies
between 75 and 105 Å. This is well beyond the limit of resolution of the light
microscope. The membrane is composed principally of protein and phospholipid
and there is evidence that these components may be arranged in a highly ordered
fashion. In one widely accepted model of membrane structure, shown in Figure 1,
the lipid and protein molecules are arranged in a sandwich. Two protein monolayers form the outer and inner surfaces of the membrane and the lipid molecules
in a double layer form the 'filling' of the sandwich. Each lipid molecule has a hydro-
philic polar end associated with the protein component of the membrane and a
hydrophobic non-polar end extending inwards at right angles to the plane of the
membrane. The lipid molecules are therefore parallel to each other. In the centre
of the membrane a narrow gap is believed to exist between the opposed non-polar
ends of these two layers of lipid molecules.

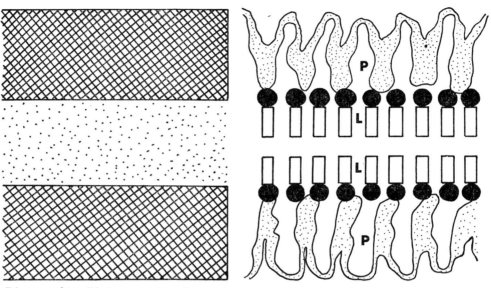

Diagram of possible 'unit membrane' structure

FIG. 1 On the left is a representation of the trilaminar appearance of a biological membrane as seen by transmission electron microscopy. The two external dense laminae are separated by a pale interspace. On the right is a possible molecular interpretation of membrane structure. The double layer of lipid molecules, L, with their non-polar ends opposed is thought to make up the interspace. The polar ends of these molecules, shown as black circles, and the associated protein layers, P, form the external dense laminae on the electron micrograph.

This lipo-protein molecular sandwich is often termed the *unit membrane*, a name which implies that all biological membranes despite differences in functional and metabolic properties have this same fundamental molecular pattern. The unit membrane theory also states that the lipo-protein unit is the smallest molecular grouping which has the properties of a membrane. An alternative interpretation suggests that biological membranes may be a mosaic of globular units or micelles, each consisting of a lipid core with a surface protein monolayer. Experimental studies of membrane permeability have led to the belief that minute pores must exist in the membrane, each less than 10 Å in diameter, but the nature of these pores and the arrangement of the molecules around them is not clear.

Despite argument about the detailed molecular pattern of the protein and lipid components of different membranes, electron microscopic observations suggest an underlying similarity of molecular architecture. In sections of fixed tissue, membranes appear at low magnification as thin dense lines, their density indicating their affinity for the heavy metal atoms used in fixation and staining. At higher magnification, if the resolution is adequate, membranes have a three layered or *trilaminar* structure, consisting of two dense outer laminae each about 25 Å thick separated by a pale interspace also of 25 Å. In the lipo-protein lamellar hypothesis the interspace represents the non-polar lipid component of the membrane and the dense outer laminae correspond to the polar ends of the molecules with their associated protein monolayers.

Although the trilaminar pattern can usually be demonstrated in any biological membrane under ideal conditions, the exact dimensions of the different components vary considerably and additional structural features and variations have been described. The cell membrane may for example be of different thickness at different points in the same cell. At the sides and base of the intestinal epithelial cell the membrane has a total thickness of about 85 Å, while at the absorptive surface at the apex of the cell it is increased in thickness to about 105 Å and is very easily resolved into its three component laminae. The pale interspace is about 40 Å wide and the inner dense lamina often seems thicker and more dense than the outer, giving the membrane an asymmetrical appearance. Within the cytoplasm, the membranes of the endoplasmic reticulum are thinner than those of the Golgi apparatus. To a certain extent, however, the dimensions and fine structure of a membrane as seen in the electron microscope are dependent upon the techniques used in specimen preparation, since the membrane is only made clearly visible by the contrast enhancing effect of the heavy metals deposited on the molecular layers during fixation and staining. For this reason detailed theories of membrane composition cannot be based on results obtained from microscopy alone. Full investigations of membrane architecture must also make use of a wide variety of physical and chemical techniques.

MEMBRANE FUNCTIONS

The function of any membrane has two interrelated aspects, the structural and the metabolic. On the one hand, membranes are essential in the construction of the cell organelles and in the partitioning of the cell from its neighbours and from the extracellular environment. On the other hand, membranes are adapted to serve many different biochemical functions through the location of enzymes at their surfaces. It is possible that some of the differences in thickness and symmetry of the trilaminar structure in different membranes may be due to the presence of protein and lipid components with different functional properties. Enzyme molecules may be incorporated in the protein layers of the membrane resulting in built-in biochemical specialisation. In this way metabolic and structural roles can be uniquely combined. Since many of the biochemical reactions which take place at membrane surfaces are limited principally by the total area of the membrane surface concerned rather than by the availability of substrates, the presence of extensive membrane systems is a common accompaniment of specialised cell function.

Efficiency of function in specialised cells is gained by accommodating the maximum active surface within the minimum volume. The granular endoplasmic reticulum, the membrane system concerned in protein secretion, is elaborately developed in protein-secreting cells. The mitochondria, membrane-based sites of energy production, are large and numerous and have an elaborate internal structure in cells which consume large amounts of energy in their metabolic processes. Specialisations of the cell surface, particularly in epithelial cells, may be related to their specific functions. The numerous finger-like microvilli at the luminal surface of the intestinal epithelial cell can increase the surface area of the cell by a factor of twenty or more. This large surface probably provides attachments for a mosaic of enzymes related to the terminal digestion of various foodstuffs and facilitates absorption. The functional specialisation of this membrane surface may account for its unusual thickness. An increase in the basal surface area of epithelial cells produced by deep infolding of the cell membrane is found in situations where there is marked ionic transport across the cell. Examples of this specialisation are seen in the cells of the kidney tubules and the acid-secreting cells of the stomach.

The surface membrane of the cell, particularly important in its structural role in maintaining the integrity of the cytoplasm, is also able to control or influence the passage of all materials into and out of the cell. Among the enzymes associated with the cell surface is adenosine triphosphatase (ATP-ase) which releases energy from ATP for use in metabolic processes. This enzyme may provide the energy required to drive the 'sodium pump', an ionic transport system which is located in the cell membrane and is thought to be essential for the maintenance of the normal membrane potential of the cell. Thus it seems likely that the metabolic role of the cell membrane may be as important as its structural role, on account of its ability to influence the internal environment of the cell by its metabolic activity.

Membranes also play a part in the isolation of materials within the cell. The metabolically active membranes of the granular endoplasmic reticulum form a storage space and channels for the isolation and transport of newly synthesised protein secretions. The membrane which surrounds each secretion granule released from the Golgi apparatus not only isolates the granule prior to its discharge but may also modify its contents by continued enzymatic action. The membrane surrounding the lysosome is apparently able to resist the action of the powerful hydrolytic enzymes of the lysosome. The structural role of this membrane is of particular importance, since rupture of the lysosome leads to cell damage and even cell death. The many variations of structure and function in different membrane systems are an indication of the flexibility which has made the basic membrane pattern so indispensable to all types of cell.

PINOCYTOSIS
(Plates 13b and 17a are relevant to this section.)

Pinocytosis (cell drinking), a process described by light microscopy some years ago, is another means whereby the cell membrane can regulate the passage of substances into the cell. This activity is best observed in living cells in tissue culture, particularly the macrophage and the amoeba. At some point on the cell surface the membrane becomes deeply invaginated into the cytoplasm to form a membrane-lined pinocytosis channel, containing extracellular fluid and dissolved materials still in continuity with the exterior of the cell. The walls at the neck of this channel then fuse, leaving its blind end as a pinocytotic vacuole isolated in the substance of the cell. The wall of the vacuole is still formed by the membrane which was originally part of the cell surface. The contents of this vacuole, although now surrounded by cytoplasm, are not free to enter the cytoplasm and can still in one sense be considered to lie 'outside' the cell as long as the membrane limiting the vacuole remains intact. When certain substances known as inducers are dissolved in the extra-cellular fluid they are capable of causing an increase in the rate and volume of pino-cytotic uptake.

Pinocytosis is related to phagocytosis, described many years previously. In both of these processes the vacuole formed, with its fluid or particulate contents, serves the function of the 'intestinal tract' of the cell. Digestive enzymes can be released into the vacuole for the hydrolysis of its contents, which can then be absorbed into the cell through its intact limiting membrane. The uptake of material by specialised cells forms an important part of the defence mechanism of the body against foreign material.

Although true pinocytosis is a phenomenon easily visible at light microscopic level, a similar process is thought to take place at the level of fine structure. Appearances taken to be suggestive of *micropinocytosis* include small indentations or invaginations of the cell membrane a few hundred Ångstroms in diameter, often narrowed at their necks to form flask-shaped vesicles known as *caveolae*. In the cytoplasm below the cell surface small circular membrane profiles of similar dimensions are often seen, perhaps formed by detachment of the flask-shaped vesicles from the surface membrane. In some cases the surface of the cell appears to have a receptor mechanism which promotes the uptake of specific molecules by selective micropinocytosis. The outer surface of the vesicle forming at the cell surface appears thickened by an external fuzzy coating. Isolated 'coated' or fuzzy vesicles can be seen within the cytoplasm in macrophages and at times in endothelial cells. A similar process may assist the uptake of ferritin by red blood cell precursors.

Appearances suggestive of micropinocytosis are common in the endothelial lining cells of capillaries, where the process could assist transfer between tissues and cir-culating blood (Fig. 2). Similar vesicles are seen at the surface of smooth muscle

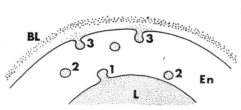

Diagram of micropinocytosis

FIG. 2 Part of a capillary endothelial cell, En, is shown with its external basal lamina, BL, and the capillary lumen, L. Micropinocytosis could assist transport across the capillary wall by the formation of a vesicle or caveola, 1, which would contain material from the lumen. The vesicle could pinch off, 2, and pass to the other surface of the cell where it could fuse with the surface membrane and release its contents, 3. The morphological appearance of micropinocytosis could signify transport in either direction.

cells, where their function is obscure. Micropinocytotic vesicles can sometimes be seen at the surface of intestinal cells between the bases of adjacent microvilli. Particulate fat from the intestinal lumen appears on occasions to become enclosed within such vesicles and it has been suggested that micropinocytosis might provide a pathway for fat absorption. There is now dispute concerning the proportion of fat absorbed in this way.

Since it is not yet possible to study living cells satisfactorily by electron microscopy, the process of micropinocytosis cannot be directly observed as pinocytosis was. The course of events taking place during a dynamic process must be inferred from the appearance of fixed material. Each micrograph represents a single moment of time in a single cell and a tentative composite picture must be constructed from different micrographs. The mere presence of vesicles as described above is only presumptive evidence for the existence of a dynamic process and gives no indication of the direction in which the process may be taking place. Additional experimental evidence is necessary to determine the direction and the quantitative significance of micropinocytosis in different sites.

The existence of the process of pinocytosis raises the question of the site of origin of membranes in the cell. If pinocytotic vacuoles can be constantly produced by invagination of the cell membrane and can then be dispersed in the cytoplasm, it is clear that there must be a constant resynthesis of membrane material to make good the loss at the cell surface. The site of membrane production in the cell is not known. A membrane may perhaps be able to expand in extent by the addition of material at any point, a diffuse synthesis of new membrane material taking place when there is a demand. Alternatively there may be a single main source of membrane material in the cell, perhaps the Golgi apparatus. There is no firm evidence to support either view. It has been suggested that the membrane systems in the cell may at times be able to become continuous with each other. The outer membrane of the nuclear envelope is at times continuous with the endoplasmic reticulum and the nuclear envelope appears to re-form from fragments of endoplasmic reticulum following mitosis. There may at times be continuity between granular and smooth endoplasmic reticulum. There is also evidence for functional connections between the endoplasmic reticulum and the surface membrane, although this remains to be conclusively demonstrated in animal cells. It is still not clear, in view of the structural and functional differences between different membranes, to what extent membrane interchange normally takes place.

CELL CONTACT AND ADHESION
(Plates 2, 3, 4, 25 and 31 are relevant to this section.)
When cells lie in contact, membrane specialisations can often be seen with the electron microscope at points along their contact surfaces. Different tissues vary in the extent to which fixed contacts occur between their cells. There are no fixed contacts between circulating blood cells and in other forms of connective tissue abundant intercellular material may separate the cells. In epithelial tissues on the other hand where the intercellular space is small, adhesion between cells must often be strong for functional reasons and specialised areas of contact between adjacent cells are common. Contact specialisations are best studied in a simple columnar epithelium, such as the intestinal epithelium, sectioned in the long axis of the cells, where the base of each cell lies close to the underlying connective tissue and the free apical surface forms the absorptive surface of the intestine. The lateral surface of the cell extending from the base to the apex is the contact surface. In stratified

epithelia with several layers of cells, such as skin, the cells in the middle layers are in contact with their neighbours over their entire surface.

The greater part of the contact surface is without obvious structural specialisation. The external surface of each cell is defined by its membrane, which appears at low resolution as a single thin dense line. At unspecialised areas the membranes of adjacent cells lie parallel to each other, separated by an interspace measuring between 150 and 200 Å. Two images associated with membranes can at times be confused. The first is the electron microscopic appearance of two dense parallel membranes of adjacent cells at low magnification, with an intervening pale intercellular space. The second is the trilaminar appearance of a single membrane at high magnification, in which the two dense components of the 'unit membrane' and their pale interspace can be seen. Confusion will not arise if the magnification of the micrograph is carefully noted.

An area of membrane specialisation always present at the apical or distal end of the contact surface in the intestinal epithelium is seen by light microscopy as the terminal bar. Three areas of distinctive fine structural specialisation, which together form the *junctional complex*, make up the terminal bar. Proximal to the junctional complex, in other words towards the base of the cell, other specialised areas take the form of small plaques. The most apical part of the junctional complex, lying next to the surface of the epithelium, is described as the *tight junction, close junction* or *zonula occludens*. The term zonula (girdle) is used to emphasise the fact that this specialised area encircles the apex of the cell forming an unbroken attachment to adjacent cells. At the zonula occludens the normal 150 to 200 Å space between adjacent cells is obliterated. Over this limited area there is fusion between the outer components of the trilaminar membrane at the surface of each of the contacting cells.

There is evidence that permeability to ions between cells, normally low at contact surfaces, is greatly increased at such junctions in certain epithelia. The ready passage of ions and the low electrical resistance which this permeability provides between cells allows a closer functional association than has previously been thought possible. It has been suggested that in certain circumstances the basic unit of ionic internal environment may not be the individual epithelial cell, but the entire epithelial group. Close junctions are also present in other tissues such as smooth muscle and they form part of the cell contact area at the intercalated disc of cardiac muscle. Their function in these sites may again be as areas of permeability to ions and low electrical resistance and they may therefore promote the spread of surface excitation from cell to cell.

The second specialised area of the junctional complex is sometimes known as the *zonula adhaerens*, or *intermediate junction*. It lies on the proximal side of the close junction and again forms a continuous girdle or zonule around the cell. At the zonula adhaerens the membranes of adjacent cells are no longer fused, but are separated by an interspace of about 200 Å. The cytoplasmic or inner surface of each membrane at this area is thickened, with fine fibrils forming a dense feltwork and extending from their attachments at the membrane into the adjacent cytoplasm. Normally no dense material is seen between the cells at this point, although the presence has been suggested of some form of intercellular cement, perhaps mucopolysaccharide in nature, which is not seen with conventional electron microscopic techniques. As its name suggests, the zonula adhaerens seems designed to promote cell adhesion.

At a variable distance below, or proximal to, these two components the third specialisation, the *macula adhaerens*, or *desmosome* is found. Desmosomes form discrete plaques like buttons on the contact surfaces rather than a continuous girdle. A desmosome is usually present at some point close to the junctional complex, but they are also found at intervals elsewhere on the contact surfaces of epithelial cells. The detailed structure of the desmosome varies in different species and different cell types. There is a space between the adjacent cell membranes of between 200 and 250 Å, occupied to a variable extent by intervening dense material. This extracellular component of the desmosome may represent a condensation of cement or cell coat material. As in the zonula adhaerens there is a thickening on the cytoplasmic sides of the desmosome to which cytoplasmic fibrils are attached.

The desmosome is primarily concerned with cell adhesion. When the contact surfaces of cells are separated by force the desmosomes are the last points at which adhesion is maintained. The cytoplasmic fibrils which are associated with the desmosome are often seen linking adjacent desmosomes. They probably represent a fibrillar structural framework within the cell which helps to maintain its shape and rigidity in relation to its neighbours. In the epidermis the bundles of fibrils associated with the demosomes are the tonofibrils seen by the light microscope.

The principal function of the junctional complex is mechanical attachment between cells. The apical surface of the cell is often exposed to external deforming forces which are resisted by the junctional complex in order to preserve the coherence and integrity of the epithelial surface. The junctional complex may also act as a seal at the free margin of the epithelium to ensure that the passage of materials in either direction across the epithelial surface takes place through the cells, and therefore under their control, rather than between them. The possibilities of communication between cells at the junctional complex may help to explain the co-ordination of function which exists for example between adjacent ciliated cells and may be of a general significance in other ways not yet realised.

THE CELL COAT
(*Plates 5, 25, 44a, and 47 are relevant to this section.*)
In many forms of life an elaborate external coating or shell is built up outside the true cell membrane. In plants, this coating forms the cell wall which consists principally of cellulose and provides an important structural scaffolding. A cell wall is also found in bacteria and an additional more diffuse external layer, the capsule, is often present. It now seems likely that all cells have some form of external polysaccharide-containing coat, although on microscopic examination this is present as a significant structural feature only in certain cases. The term *glycocalyx* has been proposed for this layer, which has also become known as the *cell coat*.

A clearly defined cell coat is found covering the cell membrane at the free surface of the intestinal absorptive cell in certain species including man. This layer has been called the fuzzy coat of the microvilli. At high resolution the fuzzy coat appears as fine filaments or 'antennulae microvillares' radiating out from the trilaminar membrane. They bridge the narrow spaces between the microvilli and extend to about a tenth of a micron beyond their tips. The fuzzy coat has been shown to contain an acid mucopolysaccharide component attached to the cell membrane and is not merely a mucus film on the intestinal surface. Investigations using auto-radiography have revealed that its mucopolysaccharide component is synthesised continuously by the epithelial cells themselves and that there is rapid turnover of this component of the coat. In the intestine, the fuzzy coat is visible on electron microscopy only at the apical surface of the absorptive cell, but it is thought that the glycocalyx may persist in some tenuous form, not necessarily demonstrable by the present techniques of electron microscopy, at the sides and base of the cell as well. A layer of this type with no affinity for the usual electron stains may occupy the apparent 150 Å space which is found between adjacent cell membranes at the unspecialised parts of their contact surfaces.

Several functions can be suggested for the fuzzy coat. It may form a protective layer, preventing possible physical and chemical damage to the cell membrane and acting as a buffer against sudden changes in the external environment. The intestinal epithelium is exposed to contact with food particles, digestive enzymes, gastric acid and bacteria in the lumen. The surface layer might well be important, as a defence against injuries which could lead to disease.

The glycocalyx in general may present a selective barrier to diffusion, allowing some substances and not others to come into direct contact with the cell membrane. This function would have general significance in relation to cell metabolism and special importance at an absorptive surface. The receptor mechanism which controls the selective form of pinocytosis could perhaps be situated in the cell coat. Enzymes may be located within the glycocalyx with functions relevant to surface activity. Through its selective permeability and enzyme activity the cell coat may help to control the composition of the 'micro-environment' which lies immediately external to the true cell membrane.

Finally, the glycocalyx may be important in determining the relationships between cells. Association between cells of similar types may be promoted by recognition of distinctive surface properties carried in the cell coat. The normal uptake of lymphocytes by endothelial cells of venules in lymph nodes has been shown to be stopped by damaging the surface coating of the lymphocyte by the action of glycosidases which do not harm the cell in other ways. It is possible that disturbances of cell association in malignant disease could be related to abnormalities of the cell coat.

BASEMENT MEMBRANE
(*Plates* 6, 17, 18, 20, 24, 28a, 29, 34a, 35, 39 *and* 40 *are relevant to this section.*)
With the recent increase in our knowledge of fine structure, the word 'membrane' has acquired a more specific definition than in the past. The light microscopic

'basement membrane' is not a membrane in the fine structural sense, but represents the interface at the base of an epithelium where there is an apparent condensation of connective tissue components.

When the epithelial basement membrane of the light microscope is examined by electron microscopy, several distinct structures can be identified. First there is the true cell membrane at the base of the epithelial cell, which appears trilaminar on high resolution microscopy. Below the base of the epithelial cell a clear zone about 300 Å wide separates the cell from an underlying homogeneous lamina of medium density, composed of diffuse flocculent or perhaps filamentous material with no organised fine structure. The latter is known as the *lamina densa*, or *basal lamina* of the epithelium. It is 300 to 700 Å thick, much thicker than a true membrane structure. It is however too thin to be visible on light microscopy and does not correspond directly with the 'basement membrane' of the histologist. Deep to the lamina densa lie collagen fibres and thin processes of fibroblasts. The collagen fibres often run parallel to the basal lamina and may seem at times to merge with it. Thus the image of the 'basement membrane' as seen by the light microscope is produced by a combination of the images of the cell membrane, the lamina densa, the collagen fibres and the connective tissue cell processes, along with any components of the ground substance which may be able to react with the histological stain being used. The term 'basement membrane' is therefore imprecise in descriptions of fine structure.

The basal lamina varies greatly in appearance in different situations. It is sometimes delicate and even apparently discontinuous, at other times thick and coherent. Descemet's membrane in the cornea, among the thickest of basal laminae, is thick enough to be seen distinctly with the light microscope. Although the basal lamina is usually closely applied to the epithelium, it does not follow infoldings at the cell base such as are seen in kidney tubule cells, but usually forms a flat shelf upon which the base of the cell, however complex, can rest as a whole. When cells are shed from the epithelial surface, the gap which they leave is rapidly filled by migration of neighbouring cells. The even surface of the basal lamina may be of importance in this process of orderly cell replacement. Since the basal lamina must be crossed by all substances which enter or leave the base of the cell, it has a potential significance in the context of epithelial metabolism.

The 'basement membrane' of an epithelium was thought by histologists to arise by condensation of the connective tissue ground substance around the base of the epithelial cells. It has recently been shown that there are antigenic cross reactions demonstrable by the fluorescent antibody technique between epithelial basement membranes in different sites and species showing the presence of a common 'epithelial' antigen. It is thought that this material may be located in the basal lamina. In certain cases it has been shown that the basal lamina is produced by the related epithelial cells and it seems likely that this may be a general rule. In this and possibly in other respects the basal lamina and the cell coat may be related.

Basal laminae are not only found in relation to epithelial cells. Diffuse *external laminae* with the same ultrastructural features are found close to the surface of muscle cells, Schwann cells of peripheral nerve and capillary endothelium. Such layers may combine some of the features of the cell coat with those of the basal lamina. They are antigenically different from epithelial basement membranes.

CHAPTER 3

Structure and Function of the Cell Components

NUCLEUS

(*Plates 1, 10, 15, 16, and 24 are relevant to this section.*)

MORPHOLOGY In fixed and stained preparations for light microscopy, the *nucleus* is often the most distinctive feature of the cell, staining particularly well with blue basic dyes such as haematoxylin. The patches of densely-staining chromatin are often distributed around the margins of the nucleus and are separated by paler areas. The chromatin pattern often presents characteristic appearances which vary from cell to cell. In some nuclei the chromatin pattern is diffuse, in others clumped. In the plasma cell the chromatin stains heavily and forms a characteristic 'cart-wheel' pattern. The main component of the nucleus is deoxyribonucleic acid (DNA) which exists in combination with a protein component. The DNA of the nucleus is the material which determines heredity. The Feulgen reaction, a histochemical test used in light microscopy, stains DNA specifically.

This description applies to the nucleus in a cell which is not dividing. Such a cell is said to be in interphase. The common form of cell division, known as *mitosis*, involves important changes in nuclear morphology. During prophase, the initial phase of mitosis, there is a progressive change in the pattern of the nuclear material resulting in the formation of clearly visible basophilic threads known as the chromosomes, in which the DNA is located. The nuclear envelope, which separates the nucleus from the cytoplasm, breaks down then. In the next stage of mitosis, called metaphase, the chromosomes line up together across the centre of the cell. A refractile double cone called the spindle is present in the cell at this stage and the chromosomes lie in the equator of the spindle. During anaphase the chromosomes draw apart to the opposite poles of the spindle. In the final stage, known as telophase, the two chromosome masses form two daughter nuclei, the chromosomes losing their separate identity, although they must still exist in some less obvious form in the interphase nucleus. The nuclear envelope re-forms around each chromosome mass and the cytoplasm splits to form two daughter cells. This activity of the nucleus is one of the most interesting aspects of cell behaviour at the light microscope level.

In the interphase nucleus the patches of chromatin recognised by light microscopy can be identified in the electron micrograph. Chromatin has a granular dense appearance on electron microscopy, the patches forming discrete clumps within the pale background of nuclear ground substance. Chromatin patches are often distributed peripherally in the nucleus and lie in contact with the inner nuclear membrane. Chromatin is also found in scattered islands elsewhere in the nucleus and at times in close association with the nucleolus. The material of the chromatin clumps shows a fine dense granularity, the individual particles being at least

50 Å in diameter. At times appearances are seen which suggest the existence of coiled threads of dense material sectioned in different planes. However, the relationships between dense areas in chromatin and the presence of DNA or nucleoprotein is not yet clear. Between the chromatin clumps extends the relatively pale, finely granular ground substance, in which there is little obvious interelationship between the dense particles which are seen. It is difficult with material of this nature to detect three dimensional patterns of arrangement involving small structural units. Scattered throughout the nucleus are larger discrete dense granules, about 150 Å in diameter, with a slightly angular appearance. These particles are structurally similar to the ribosomes of the cytoplasm and to particles which are found in the nucleolus. It is possible that these may be newly-formed ribosomes, synthesised in the nucleolus, which have not yet passed to the cytoplasm.

It is also difficult to read meaning into the fine structural image of the mitotic chromosome, since once again the correlation between the areas of electron density and the chemical constituents of the chromosome is not fully understood. The chromosome is an irregular mass of dense material which is not partitioned in any way from the surrounding ground substance of the cell. Different levels of organisation have been described by electron microscopy. Strings of dense particles interpreted as coarse aggregates of DNA and nucleoprotein are seen, as well as fine spirals and chains of dense material at times as small as 20 Å in diameter, the approximate molecular diameter of DNA. In thin sections used for electron microscopy only a small portion of each chromosome is normally found lying in the plane of section. The examination of the number and form of the chromosomes is possible only when they are spread out in a histological preparation for light microscopy, so that for many problems concerning the chromosomes the light microscope can give more valuable information than the electron microscope.

FUNCTION The nucleus is not only essential for cell division, since a cell may live for many years without dividing, but is also important in the daily control of the cell. Any cell deprived of its nucleus will soon die. Protozoa so treated lose their motility and their power to feed. Once the red blood corpuscle of the mammal loses its nucleus during maturation it not only becomes unable to reproduce itself, but its potential lifespan becomes limited. It is now clear that the nucleus directs the activity of the cell according to genetic information carried in the nuclear DNA.

The molecule of DNA forms a long chain arranged in a double stranded spiral, the two strands of which are complementary. The precise sequence in the chain of the bases in the nucleotides, the molecular subunits of the DNA molecule, forms a genetic 'code' in which the information essential for heredity is carried. The molecule of DNA has two important properties which make it particularly suitable as a carrier for the genetic code. The first feature is its ability to reproduce itself under suitable conditions. Each molecule of DNA splits longitudinally into two complementary halves. Each half is then a template for the reconstruction of a new molecule identical to the original. The molecule is constructed in such a way that mistakes during replication are extremely rare. The second feature of DNA is its stability. The sequence of bases, and thus the genetic code, remain exactly the same irrespective of the number of cell divisions or of the passage of time. The only way in which the code is altered is by physical or chemical damage to the DNA molecule.

Shortly before mitosis, the DNA of the nucleus doubles itself in a brief synthetic phase. The subsequent cell division results in the distribution of the full complement of genetic information to each of the daughter cells. In this way the genetic code is carried intact to every cell of the body. The germ cells, the sperm and the ovum, are the only exceptions. They are formed by a specialised type of cell division, called meiosis, in which there is a halving of the normal genetic information. Each germ cell has half of the full genetic code, and is described as 'haploid' on this account. They are the only haploid cells in the body. When the germ cells fuse, the zygote which is formed once more has the full complement of genetic material and is described as 'diploid'. The subsequent mitotic divisions of the zygote produce the cells of the new individual, all of which have the same genetic make-up as the zygote. If at any stage the sequence of bases is altered in the DNA molecule a mutation is said to have occurred. The alteration may be so slight as to be un-noticed, or so severe as to cause the death of the cell or prevent it from ever dividing successfully. Since the mutation involves an alteration in the DNA molecule, it is a permanent change in the genetic code of the cell and is transmitted to all the progeny of the mutant cell, so long as mitosis is still possible following the injury.

The nucleus exerts control over the day-to-day activities of the cell by determining through the genetic code the nature of every protein molecule synthesised by the cell. Since this includes the enzymes necessary for each step in metabolism this control by the nucleus extends to all forms of cellular activity. However, in any cell of the body, only part of the nuclear DNA will be concerned with the active control of cell function at any given time. The remaining part, containing information in the genetic code which is not required by the cell, must remain inert. The nucleoprotein which is associated with nuclear DNA may play a part in the control of the expression of the genetic code. It is now believed that the dense patches of chromatin present in the interphase nucleus represent the inert, coiled portions of the chromosomes. This material is sometimes referred to as heterochromatin. The paler areas of the nuclear ground substance, termed euchromatin, probably contain the extended, metabolically active portions of the chromosomes which are engaged in the synthesis of specific messenger ribonucleic acid (RNA).

In the past, the nucleus has often been more intensively studied than the cytoplasm by cytologists using the light microscope. This was partly because of the striking activity of the nucleus during cell division and partly because of the difficulties of resolving the finer cytoplasmic details. To a certain extent the electron microscope has reversed this emphasis in cytology. Since all nuclei contain the same essential biochemical mechanism for the direction of the cell, it is not surprising that the nucleus has been relatively unrewarding to electron microscopic study. Differences between the nuclei of different cells are expressed in differences at molecular level rather than at the level of electron microscopic fine structure.

B

NUCLEOLUS

(Plate 16 is relevant to this section.)

MORPHOLOGY The nucleus normally contains one or more small discrete structures clearly seen on light microscopy, termed *nucleoli*. A nucleolus is, of course, not always seen in thin sections of the nucleus, since the plane of section may not pass through that part of the nucleus in which it lies. The nucleolus is basophilic and has been shown to contain a substantial proportion of RNA, by cytochemical techniques and by biochemical analysis. By light microscopy two components of the nucleolus have been described, a coiled coarse skein of material called the nucleolonema and a paler substance between the loops of the nucleolonema which is termed the pars amorpha. Patches of chromatin are often associated with the rim of the nucleolus.

On electron microscopy the nucleolus is readily identified, if the plane of section permits, by its high electron density. The nucleolus is not separated from the substance of the nucleus by any definite specialisation such as a membrane. The two components of the nucleolus, the nucleolonema and the pars amorpha, can be identified. The nucleolonema, the denser part of the nucleolus, may have a coiled structure, or may at times best be represented as a sponge-like component. The pars amorpha, less dense, fills the interstices between the twisted components of the nucleolonema. The main fine structural subunit of the nucleolonema is a slightly angular dense particle, structurally similar to the ribosomes seen in the cytoplasm. Similar particles are found elsewhere in the nucleus. In addition, thin fibrils less than 100 Å in diameter have been described within the nucleolus in association with these granules. The granules often lie in groups, densely packed, but may also be more loosely arranged. No more detailed three dimensional pattern has yet been clearly recognised. The pars amorpha often appears similar to the nuclear ground substance in its electron microscopic features, containing only fine granules with few obvious interrelations.

FUNCTION The nucleolus is an essential link in the chain of communication between the nucleus and the cytoplasm. The initial step in the biosynthesis of protein is the formation of messenger RNA, a molecule which contains the detailed information originally locked in the genetic code of DNA. This information is necessary for the linking of amino acids in the correct sequence to form individual protein molecules, a process which takes place on the ribosomes in the cytoplasm. It is thought that the granules of the nucleolonema may in fact be newly synthesised ribosome subunits, which pass out of the nucleolus into the nuclear substance, and finally into the cytoplasm, combining at some point with the messenger RNA already formed by the DNA of the nucleus. The messenger RNA can then 'instruct' the ribosomes to synthesise a specific protein. In this way the nucleus can direct cell function through its messenger RNA, perhaps using the ribosomes, synthesised by the nucleolus, as its means of transport into the cytoplasm where protein synthesis takes place. It is obvious that the passage of material from nucleus to cytoplasm must be of great importance in the communication of genetic information to the cell, but it is not yet clear along what physical pathways this transfer takes place. The importance of the nucleolus in protein synthesis is emphasised by the prominence of the nucleolus in cells with marked activity of this type, such as tumour cells engaged in rapid growth.

NUCLEAR ENVELOPE

(*Plates* I, 12a, 15, 16 *and* 24 *are relevant to this section.*)
MORPHOLOGY The partition which separates the nucleus from the cytoplasm is a more complex structure than the surface membrane of the cell and is known as the *nuclear envelope*. It is composed of two separate membranes, each of which shows the characteristic trilaminar membrane structure at high resolution. The inner nuclear membrane forms the limit of the nuclear contents and is separated by a gap of about 500 Å from the outer nuclear membrane. The width of this gap varies in different cells and even at different points around a single nucleus. This narrow cavity surrounding the nucleus is called the *perinuclear cisterna*. On one side the cisterna is in contact with the chromatin clumped on the inner surface of the inner nuclear membrane and on the other side the outer nuclear membrane lies in contact with the cytoplasmic ground substance.

In cells with a prominent granular endoplasmic reticulum, continuity may be seen between the outer nuclear membrane and the membranes enclosing one of the cytoplasmic cisternae. The ribosomes, many of which are attached to the cytoplasmic surfaces of the membranes of the granular endoplasmic reticulum, are also commonly seen on the outer surface of the outer nuclear membrane. As a result of this relationship, there is a potential intercommunication between the cavities of the endoplasmic reticulum and the perinuclear cisterna. Since it is known from light microscopic examination of the living cell that its interior is in a state of constant movement it seems likely that such connections are being constantly broken and re-formed.

Nuclear pores form an interesting and important specialisation of the nuclear envelope. These pores, usually from 500 to 700 Å in diameter, are more common in some cells than in others, the area occupied by them ranging from 5 per cent to 30 per cent of the total surface area of the nucleus. At the nuclear pore the inner and outer nuclear membranes become continuous, producing a circular area of potential communication between the nucleus and the principal phase of the cytoplasm. The pore thus formed is usually bridged by a diaphragm which has no obvious trilaminar structure and is more diffuse in appearance than a typical membrane. The diaphragm is the only apparent barrier across the nuclear pore. Around the margins of the outer aspect of the pore a dense collar or annulus is often seen projecting into the cytoplasm. When a tangential section of the nucleus is examined, the nuclear pores and their annuli can be seen in 'surface' or 'face' view. The annuli can at times be seen to consist of a ring of tiny subunits extending out from the circumference of the pore into the adjacent cytoplasm. Finally, in certain cases, a 'flange' can be seen on cross section of the nuclear pore, extending into the perinuclear cisterna in the plane of the diaphragm. The detailed morphology of the nuclear pores is subject to great variation, but the reasons for this variation are not yet clear.

FUNCTION It is now accepted that communication between nucleus and cytoplasm is important in the control of the cell by the nucleus. Since the nuclear envelope must be crossed by any molecules taking part in this interchange, it is possible that the nuclear pores play a part in effecting or controlling this biochemical communication. The extent to which the nuclear pore presents a barrier to diffusion is not known, but it is the only clear-cut structural channel between the cytoplasm and the nucleus. If the nuclear pores could be shown to exercise some form of selective control over nucleo-cytoplasmic interchange, they would become established as structures of the greatest significance.

The potential continuity of the perinuclear cisterna and the endoplasmic reticulum is underlined by other observations. During fat absorption by the columnar cells of the intestine, droplets of particulate lipid material are found in the channels provided by the endoplasmic reticulum. On occasion, however, lipid droplets have also been seen within the perinuclear cisterna. In the developing Paneth cell of the intestine, a cell with prominent granular endoplasmic reticulum, material with a distinctive crystalline appearance has been reported in both the endoplasmic reticulum and the perinuclear cisterna. Such results suggest that these cavities may share a common function, emphasising the dynamic unity of the intracellular system which they constitute. The possibility also exists that the nuclear envelope, at least in developing cells, may be a source of formation of the endoplasmic reticulum.

The behaviour of the nuclear envelope during cell division reinforces the view that it is a dynamic cell component related to the endoplasmic reticulum. In the early stages of mitosis the nuclear envelope breaks down, as the chromosomes form, becoming dispersed in the cytoplasm in the form of vesicles and membrane profiles similar to the scattered components of the endoplasmic reticulum. During most of mitosis the chromosomes lie in direct contact with the cytoplasm, without any form of intervening structural barrier. At telophase, however, when the two chromosome masses of the separating daughter cells are coalescing to form the daughter nuclei, each nucleus becomes surrounded once more with its complex envelope, apparently formed by the linking up of the scattered cytoplasmic vesicles and cisternae of the reticulum. The mechanisms responsible for the orderly breakdown and reformation of the nuclear envelope are not known.

RIBOSOMES
(*Plates* 3, 7 *and* 33 *are relevant to this section.*)
Early studies of cytoplasmic fine structure showed the presence of distinctive small dense granules which have become known as *ribosomes*. They are 150 Å in diameter and have a slightly angular profile when seen at high magnification. These granules, among the most fundamental of subcellular particles, are present in virtually every type of cell. They are found singly and in groups in the cytoplasm,

often arranged in clusters or rosettes, sometimes in spirals. Identical granules are attached to the outer surfaces of the membrane-bounded cisternae of the granular endoplasmic reticulum giving this system its characteristic morphology. The nucleolus contains large numbers of similar particles, closely packed, and they are also found in the substance of the nucleus.

The name ribosome reflects the presence in these granules of a significant proportion of ribonucleic acid (RNA). When a cell contains many ribosomes it has a characteristic cytoplasmic basophilia. Diffuse basophilia is seen when numerous ribosomes are free in the cytoplasm and organised cisternae of the endoplasmic reticulum are few, as in the lymphocyte or the intestinal crypt cell. When the ribosomes are associated with areas of organised granular endoplasmic reticulum there is a patchy basophilia which has been given different names in different cell types in the past. In the nerve cell such areas were termed Nissl bodies, which lie in scattered patches throughout the cell. In the pancreatic zymogen-secreting cell the basophilic area, often predominantly basal, was termed ergastoplasm.

The RNA of the ribosome is the site of protein synthesis in the cytoplasm. This fundamental activity is finally controlled by the genetic code of DNA. The ribosomes themselves may be synthesised in the nucleolus, but they receive their instructions for the manufacture of specific protein molecules from messenger RNA in which the genetic code is transcribed. There is now biochemical evidence that each ribosome has two subunits each of which may have a specific function. A number of ribosomes concerned in the manufacture of a single large protein molecule may be arranged together in a functional group known as a polyribosome or polysome, probably linked by the thread of messenger RNA. The rosettes and spirals of ribosomes commonly seen in the electron micrograph may represent these functional units of protein synthesis.

ENDOPLASMIC RETICULUM

MORPHOLOGY In the early days of biological electron microscopy, when crude sectioning techniques were inadequate to demonstrate detailed fine structure, cells could be examined whole, spread out on thin plastic films. This method was unsatisfactory in many ways since the specimen was not only thick, but was unsupported by an embedding medium and distorted by dehydration. However the main cell features could be seen in outline. The cytoplasm was found to contain a lace-like network extending through the interior of the cell, to which the name *endoplasmic reticulum* was given. With greater refinement of thin sectioning methods for electron microscopy, the endoplasmic reticulum appeared as a complex membrane system composed of pairs of membranes enclosing interconnecting cavities or cisternae. Although wide structural variations occur from cell to cell, the endoplasmic reticulum possesses common features which have led to its recognition as a fundamental cytoplasmic organelle. Two main types of endoplasmic reticulum are recognised on a morphological basis, *granular* and *agranular* endoplasmic reticulum. The alternative terms *rough* and *smooth* are often used.

Granular Endoplasmic Reticulum (*Plates*, 1, 7, 9b, 22 *and* 23 *are relevant to this section*.) In most cells there are cytoplasmic membranes with ribosomes attached to their outer or cytoplasmic surfaces, forming a cisternal pattern classified as endoplasmic reticulum. The presence of dense granules on these membranes has led to the use of the descriptive term granular reticulum. The membrane-bound cisternae of the reticulum form patterns of varying complexity, ranging from a few small isolated cisternae to systems with interconnections and cavities which form what appears to be a continuous phase of the cytoplasm limited by an extensive area of granule-covered membrane surface. Individual cisternae usually take the form of flattened envelopes, but may appear at different points in thin section as tubular or vesicular profiles. The outer nuclear membrane which is also studded with ribosomes appears structurally similar to the membranes of the endoplasmic reticulum with which it occasionally becomes continuous. When this occurs, the perinuclear cisterna becomes confluent with the cisternae of the endoplasmic reticulum. It is now believed that the granular endoplasmic reticulum is a dynamic system. Connections between different cisternae and between the endoplasmic reticulum and the perinuclear cisterna are thought to be continually made and broken as a result of cytoplasmic activity. This activity is halted at the moment of fixation to give the static picture shown in the electron micrograph.

The ribosomes, the site of cytoplasmic protein synthesis, are attached to the membranes in variable numbers, at times closely packed, at times more widely spaced. When a tangential section of a cisterna is obtained, a 'surface view' of the ribosomes may show their arrangement in two dimensions as opposed to the single dimension demonstrated by the normal section of the wall of a cisterna. Clusters and spirals in inter-related groups are often found in addition to randomly spaced single ribosomes and it is possible that the members of a group are linked together functionally and structurally to produce molecules of a specific type of protein.

Several lines of investigation have revealed at least one aspect of the significance of the granular endoplasmic reticulum in cellular metabolism. Zymogenic cells which secrete enzymes and plasma cells which secrete antibody protein molecules, have a pronounced cytoplasmic basophilia on light microscopy, which is due to the presence of cytoplasmic RNA in the form of ribosomes associated with a particularly well developed system of organised granular endoplasmic reticulum. Biochemical studies of homogenised pancreatic cells have shown that protein synthesis is carried out by the *microsome* fraction of the cell. This fraction, which remains after the heavier cell components have been spun down, is composed to a great extent of ribosomes and fragments of the membranes of the endoplasmic reticulum.

As a result of biochemical studies of this type, the presence of an elaborate granular endoplasmic reticulum is now an accepted indication of active protein synthesis directed for eventual use outside the cell, as in the case of digestive enzymes or antibodies. Numerous cytoplasmic ribosomes with few associated membranes suggest, on the other hand, that there is significant protein synthesis being undertaken for internal use, as in the rapidly growing or dividing cell. Similarly, when protein is to be stored in the cytoplasm indefinitely after synthesis, as haemoglobin is in the red cell precursor, organised granular endoplasmic reticulum is scarce but ribosomes are plentiful. In cells which 'export' the protein they make, the membrane system may be essential for the segregation and transport of the finished product through the cytoplasm in preparation for further processing and discharge. The potential metabolic activity of the large membrane surface of the endoplasmic reticulum suggests that secretion might be added to or selectively altered while being stored or transported in the cisternae.

SMOOTH ENDOPLASMIC RETICULUM (*Plates* 8, 11, 21a, 26 *and* 27 *are relevant to this section.*) In many cells there are membrane-bound cisternae arranged in patterns related to those of the granular reticulum but without associated ribosomes. Cytoplasmic membranes of this type are all classed, despite morphological variations, as smooth or agranular endoplasmic reticulum. Although the membranes of the granular reticulum are sometimes seen in continuity with areas free of attached ribosomes, it is more common for smooth and granular reticulum to occupy different areas of the cell. Cells containing significant proportions of smooth reticulum, especially when combined with a prominent mitochondrial population, are generally found to give an acidophilic staining reaction when examined with the light microscope.

The detailed structure of the smooth endoplasmic reticulum is extremely variable and appears to depend to a considerable extent on the method of fixation. With the osmium tetroxide solutions often used to fix tissues for electron microscopy, the smooth reticulum is discontinuous, consisting mainly of empty-looking smooth-surfaced vesicles several hundred Ångstroms in diameter. With glutaraldehyde fixation, however, the smooth reticulum may appear as narrow interconnecting tubules forming a network more like that of the granular reticulum. It may be that the vacuole form of the smooth reticulum results from fixation damage and the tubular form perhaps represents a closer approach to the original state of the living cell.

Among the cells with a prominent smooth endoplasmic reticulum are the acid-secreting gastric parietal cells of the mammal and the acid-secreting cells of other species. In certain species, variations in the functional states of these cells are accompanied by changes in fine structure, with an apparent depletion of smooth reticulum accompanying the secretion of acid. The cells of the adrenal cortex and the interstitial cells of the testis, both of which have a well developed smooth endoplasmic reticulum, share the function of steroid hormone secretion. In normal circumstances liver cells have a variable proportion of smooth reticulum which can be made to increase by the administration of barbiturate drugs which are detoxicated by enzyme activity in the liver. The smooth reticulum may also be associated with glycogen metabolism in liver cells since there is a close topographical relationship between smooth-surfaced membranes and glycogen granules. The amount of smooth endoplasmic reticulum in the intestinal absorptive cell varies, but its regular appearance suggests that it performs some metabolic function in relation to digestion or absorption. Finally, an extreme specialisation of the smooth endoplasmic reticulum is seen in cardiac and skeletal muscle. In this site, the smooth-surfaced membranes, commonly termed the sarcoplasmic reticulum, appear to be concerned with calcium binding and may link the membrane excitation of the muscle cell to the contraction of the myofibrils.

FUNCTIONAL SIGNIFICANCE OF THE ENDOPLASMIC RETICULUM It is possible that the morphological subdivision of the endoplasmic reticulum into distinct 'granular' and 'agranular' components may be misleading, since it implies a clear functional distinction which has not yet been demonstrated. Although the granular endoplasmic reticulum is concerned with protein synthesis there is no single function which can be assigned at present with equal confidence to the smooth reticulum. Indeed, the wide range of cells in which smooth-surfaced cytoplasmic membranes are prominent makes it difficult to suggest a possible functional link. Perhaps it would be more correct to regard the membranes of the endoplasmic reticulum, whether granular or smooth, as having a single common function, the formation of a favourable 'metabolic environment' which becomes adapted in different ways to serve different biochemical functions in specialised cells.

A membrane-limited cisternal system has several important functional properties. First, it effectively partitions the cytoplasm into two phases separated by a single membrane surface. In this way ideal conditions are provided for the segregation of material within a closed system in the cell. Secondly, a large area of membrane surface can be accommodated within a limited cytoplasmic volume. Surface-limited reactions can therefore be carried out with increased efficiency in the confines of the system. Thirdly, enzyme systems attached to the membranes can be disposed in their optimum concentration and spatial interrelationships between enzymes can be established which would not otherwise be possible. Fourthly, such a system of interconnecting cisternae provides channels throughout the cell which could be used for the transport of material. These properties of a simple cisternal framework can easily be adapted to different metabolic activities by the attachment of different enzyme systems to the membrane surfaces. In this way, the endoplasmic reticulum can become associated with functions as different as acid secretion, steroid metabolism, absorption, detoxication and excitation-contraction coupling in different cells.

In most cases, however, the attachment of different enzymes to the membranes of the endoplasmic reticulum does not produce any fine structural specialisation of the membrane surface as studied with present techniques. The specialisation which has taken place is at the level of molecular structure rather than microscopic structure. Only one biosynthetic system is associated with a clearly recognisable structural unit; the apparatus for protein synthesis lies in the ribosome which can be identified readily in thin sections. Thus when the 'metabolic environment' of the endoplasmic reticulum common to all cells is adapted for the particular purpose of protein secretion, the membranes become structurally as well as biochemically specialised and can be recognised by electron microscopy as the 'granular reticulum'.

The full significance of the endoplasmic reticulum in cell metabolism is not yet known and even the extent of its relationships with the other parts of the cell is far from clear. It has been suggested that the endoplasmic reticulum may connect directly with the exterior of the cell, but in the animal cell this can occur only rarely, if at all. Evidence for communication between the endoplasmic reticulum and the Golgi apparatus is more convincing. It is possible that the endoplasmic reticulum may be concerned with bio-electrical phenomena.

GOLGI APPARATUS

(*Plates* 1, 9, 10, 22 *and* 33 *are relevant to this section.*)

MORPHOLOGY When cells are impregnated under suitable conditions with silver or osmium solutions a structural network is demonstrated by deposition of the metal in specific parts of the cytoplasm. Since this reticular system was first described by Golgi, it became known as the *Golgi apparatus* or *Golgi complex*. The significance of the Golgi apparatus became the subject of a prolonged cytological dispute. From observations with the electron microscope it is clear that the Golgi apparatus is a membrane system which forms a distinct structural component of the cytoplasm in which osmium is deposited selectively under the conditions of the classical Golgi impregnation. In the thin sections used for electron microscopy, the full three dimensional nature of the network is not immediately appreciated since its interconnecting strands can appear as isolated units at different points in the cell. This is especially true of cells such as the neurone in which the elaborate Golgi apparatus often surrounds the nucleus. Epithelial cells of columnar or cuboidal shape are said to be polarised, having a definite apical and basal surface. In these cells the Golgi apparatus generally lies at the apical pole of the nucleus occupying an oval or a horseshoe-shaped area of the cytoplasm in thin sections. The position and morphology of the Golgi apparatus is relatively constant in a given cell type. It is small in muscle cells and lymphocytes but is well developed in the goblet cell and the secretory cells of the exocrine pancreas and seminal vesicles, as well as in the absorptive cells of the intestine.

The three principal components forming the Golgi apparatus as seen by electron microscopy are membrane lamellae, dilated membrane-bound vacuoles and small vesicles (Fig. 8). The membranes of the Golgi apparatus often appear slightly thicker than those of the endoplasmic reticulum, and the usual trilaminar membrane structure can be made out at high resolution. The membrane lamellae which may make up most of the apparatus are paired, each pair forming a single closed sac, the walls of which can lie close together or widely separated. The interior of this sac is usually pale in an electron micrograph. The Golgi vacuoles often appear to consist of the dilated ends of such sacs and are thus continuous with the spaces between the paired membrane lamellae. It is usual for several typical Golgi sacs to lie in parallel arrangement, the separation of adjacent sacs often being less than the width of each individual sac. In this way the lamellar structure of the Golgi apparatus is built up. Around these main components of the apparatus there lies a population of small membrane-bound vesicles, each a few hundred Ångstroms in diameter. These Golgi vesicles may be very numerous and at times seem to contain material of appreciable density. The Golgi apparatus can be distinguished from the endoplasmic reticulum by its circumscribed appearance, its position in the cell, the thickness of its component membranes and the distinctive architecture of its different elements.

FUNCTION Early light microscopic studies linked the Golgi apparatus with the process of secretion in certain cells, a suggestion supported by electron microscopic evidence. In the goblet cell and the zymogenic cells of the digestive glands the Golgi apparatus is particularly complex. In some cells granules of secretion form in the cytoplasm close to the Golgi apparatus, perhaps in certain cases by the accumulation of material within components of the Golgi apparatus. These granules subsequently become detached from the apparatus and pass towards the apex of the cell where maturation, perhaps by withdrawal of fluids, may take place, and where the granules are stored prior to release. A similar relationship between secretory granules and the Golgi apparatus is also seen in certain cells with an endocrine secretory function such as the islet cells of the pancreas, the intestinal chromaffin cells and the anterior pituitary cells

The early suggestion that the Golgi apparatus may participate in secretion by concentrating or packaging material made elsewhere is supported by recent work. In protein-secreting cells such as the cells of the exocrine pancreas the major part of protein synthesis takes place in the endoplasmic reticulum, which then appears to transport this material within its cisternae to the region of the Golgi apparatus. It is possible that small vesicles filled with the newly synthesised contents of the cisternae may bud off from the membranes adjacent to the Golgi apparatus (Fig. 8). These vesicles could carry material across to the apparatus, releasing it into the cavities of the Golgi sacs. The secretory substance would then pass through the apparatus, undergoing concentration and perhaps chemical alteration, finally accumulating as a secretion granule at the far side of the apparatus. Since the final granule is surrounded by a membrane derived from the Golgi apparatus, it is possible that enzymes attached to the membrane may continue to act on the granule after its release from the apparatus. In the goblet cell it has been shown that this sequence of granule formation and release is continuous, secretion granules leaving the apparatus and passing to the cell apex at the rate of about one every two minutes.

In addition to the segregation, concentration and packaging of secretions synthesised elsewhere, the Golgi apparatus has independent synthetic functions of its own. There is evidence that the polysaccharide components of certain types of secretion may be manufactured by the Golgi apparatus where they become conjugated with a protein moiety synthesised in the endoplasmic reticulum. In the intestinal cell the Golgi apparatus has been shown to be the site of synthesis of the carbohydrate component of the fuzzy coat of the microvilli. It also takes part in the process of absorption. In certain protozoa the Golgi apparatus may be concerned with the regulation of the fluid balance of the cell.

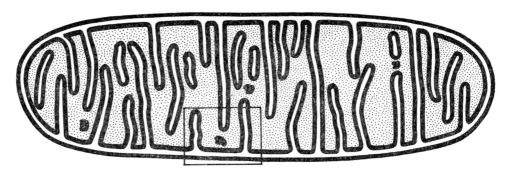

Diagram of a mitochondrion
FIG. 3 The mitochondrial membranes form a structure with complex internal shelves, drawn here as seen in sections. The small area enclosed by lines is enlarged in Figure 4.

MITOCHONDRIA

(Plates 11, 30, 31 and 33 are relevant to this section.)
MORPHOLOGY Small cytoplasmic structures in the shape of threads and granules were described by cytologists before the start of this century. These structures were thought to be intracellular organelles and they were called *mitochondria*. Rod-shaped, spherical and sinuous mitochondria were described, and wide variations in their numbers, sizes and shapes were noted in different cells. For many years, however, the significance and function of the mitochondria were disputed.

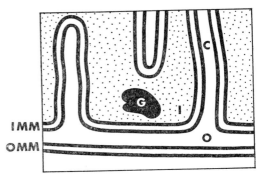

Diagram of mitochondrial structure
FIG. 4 The intramitochondrial space, I, sometimes known as the mitochondrial matrix, is shown in stipple. Dense intra-mitochondrial granules, G, are situated within this space. The trilaminar pattern of the mitochondrial membranes is shown diagrammatically. The outer mitochondrial membrane, OMM, is separated from the inner mitochondrial membrane, IMM, by the outer mitochondrial space, O, which extends into the centre of each of the cristae, C.

Electron microscopic examination has shown that mitochondria are cytoplasmic organelles of distinctive fine structure. All mitochondria are constructed of membranes arranged in a similar organised pattern shown diagrammatically in Figure 3. The membranes of the mitochondrion are thinner than the cell surface membrane, but a trilaminar pattern can be seen at high resolution. The mitochondrion is limited by the outer mitochondrial membrane, forming an unbroken boundary with the cytoplasm. Within this limiting membrane is the inner mitochondrial membrane, separated from the outer membrane by a constant narrow gap, the outer mitochondrial space, which measures approximately 80 Å in width. The inner mitochondrial membrane is thrown into shelves or *cristae*, which extend from the side wall of the mitochondrion into its centre. Each of the cristae consists of two parallel layers of the inner mitochondrial membrane separated by an extension of the outer mitochondrial space. The cristae are very variable in morphology, at times extending only a short distance into the centre, at times bridging the mitochondrion from side to side, forming virtual partitions. The space within the mitochondrion into which the cristae project is completely enclosed by the inner mitochondrial membrane. This is known as the inner mitochondrial space or the matrix of the mitochondrion. The matrix is often finely granular but varies in density, being occasionally so dense that the outer mitochondrial space at the rim of the mitochondrion and within the centres of the cristae stands out in negative contrast. In some cells there are irregular dense intramitochondrial granules situated in the matrix, with a diameter of around 500 Å. In other cases, there may be dense droplets, apparently lipid, within mitochondria, as in some cells of the adrenal cortex.

The variability in the numbers and outlines of mitochondria reported by light microscopy has been borne out by electron microscopy and although the general architecture of the mitochondrion is essentially similar in different cells, there are detailed variations in mitochondrial fine structure. In the intestinal epithelium the mitochondria are usually thin and elongated, in the liver rounded and quite large but with sparse and poorly organised cristae. Cells with a high rate of oxidative metabolism have large mitochondria with closely packed cristae. Cells with a significant role in steroid hormone metabolism such as adrenal cortical cells and the interstitial cells of the testis commonly have cristae which are tubular rather than shelf-like. The significance of these variations is unknown, although structural variations may reflect distinctive spatial arrangements of the mitochondrial enzymes. The true variations in shape and size of mitochondria are often exaggerated by the thin sections used for electron microscopy. An elongated, sinuous structure may be represented in an electron micrograph by different profiles depending on whether the plane of section cuts it obliquely, transversely or longitudinally. Tangential sections of mitochondria may present confusing images on account of the failure to resolve the mitochondrial membranes as distinct structures.

FUNCTION The biochemical techniques of differential centrifugation of cell homogenates have made it possible to isolate mitochondria for biochemical investigation. In this way it has been shown that the mitochondria contain most of the enzymes of the citric acid cycle. This is a sequence of biochemical reactions concerned with the breakdown of many of the simple molecules from which energy is obtained by the cell. The citric acid cycle is the final common pathway of catabolism for carbohydrate, fatty acids and many of the amino acids. As a molecule passes through the cycle, hydrogen atoms are removed by enzyme action and are eventually combined with oxygen to form water. During these oxidative reactions, energy which had been stored in the chemical structure of the molecule is released for use by the cell.

The energy is made available for use by a further sequence of enzyme-controlled reactions. The hydrogen atoms removed in the citric acid cycle are passed along a series of carriers which bring about their oxidation step by step through a process of electron transfer. In this way the energy of oxidation is released in small amounts and can be trapped and converted into a form which the cell can use. The energy is used to produce molecules of adenosine triphosphate (ATP) by the phosphorylation of adenosine diphosphate (ADP). The ATP molecule contains two 'high energy' phosphate bonds which provide a ready source of energy for many of the activities of the cell. In muscle contraction for example, when the ATP is broken down to ADP this energy is released and used to produce movement. The process whereby ADP is converted to ATP through the simultaneous oxidation of simple molecules is called *oxidative phosphorylation*. This is the main biochemical function of the mitochondria.

The process of electron transport is an essential factor in oxidative phosphorylation. By disrupting mitochondria, structural subunits have been isolated which apparently contain the complete electron transport chain. These enzyme assemblies have been called *electron transport particles* (ETPs) and it is estimated that from 10,000 to 50,000 of them may be present in a single mitochondrion. On the basis of negative staining techniques it has been suggested that the surfaces of the mitochondrial cristae are covered by closely packed subunits which might represent the ETPs. On theoretical grounds the more ETPs present in a mitochondrion, the greater its capacity to produce energy from simple molecules. In cells which produce large quantities of energy by oxidative phosphorylation the high 'metabolic rate' of the cell is reflected in the number, size and complexity of the mitochondria. Among the types of cell which have prominent mitochondria are the acid-secreting gastric parietal cell, the brown fat cell and the cardiac muscle cell.

Since gastric juice contains hydrochloric acid at pH 1 and the tissue fluid surrounding the base of the gastric gland cell is at a pH level greater than 7, it follows that the acid-secreting gastric parietal cell can concentrate hydrogen ions (H^+) by a factor of 1,000,000. The secretion of electrolyte against such a concentration gradient is a remarkable biochemical activity which can be accomplished only if large amounts of energy are produced by numerous large mitochondria. The heart provides the motive force for the circulation from the beginning to the end of life with a reliability greater than that of any mechanical pump. The muscle cells are in active contraction for nearly half of the lifetime of the animal and may be called on to increase the output of work by a factor of ten for prolonged periods during exercise. The energy supply for this constant activity must come from the continued oxidation of simple molecules in the mitochondria. Brown fat is particularly prominent in animals which hibernate and in the very young of warm-blooded species. The fat stored in these cells is apparently used as an energy source for the generation of heat, important for restoration of body temperature at the end of hibernation and for the maintenance of body temperature in the poorly-insulated new born. In all of these cells, the mitochondria are numerous and very large. In addition, their parallel cristae are closely packed and often cross from wall to wall of the mitochondria. These structural features provide an extensive surface area on which the electron transport particles and other enzyme systems of the mitochondria can be grouped in large numbers.

The mitochondria, the source of energy-rich ATP, are often found gathered in the particular region of the cell in which energy is being used to produce metabolic work. There are many illustrations of this tendency to minimise the gap between 'power station' and 'consumer' across which ATP must diffuse. In striated muscle the mitochondria lie in contact with the myofibrils, the contracting units of the muscle cell. In the kidney tubule cells the mitochondria lie sandwiched between complex basal infoldings of the cell membrane where large-scale transfer of water and solutes takes place (Fig. 5). The mid-piece of the spermatozoon consists of a tightly wound spiral of mitochondria around a central core of functional units which have a probable contractile function and give the sperm its motility. Mitochondria are often closely related to the granular endoplasmic reticulum (Fig. 6).

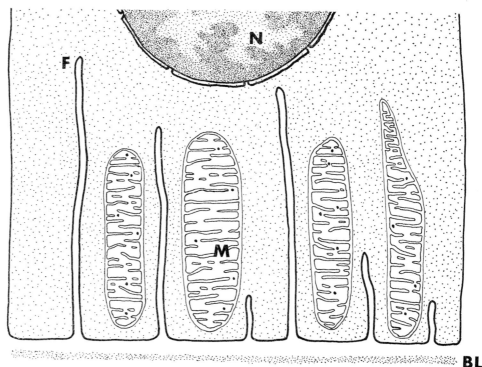

Diagram of basal infoldings of the cell membrane

FIG. 5 This pattern of cell components is seen in kidney tubule cells. The membrane at the base of the cell is thrown into folds, F, which may reach up to or beyond the level of the nucleus, N. Between the folds of cell membrane, mitochondria, M, may be situated. The basal lamina does not follow the cell membrane closely in such situations, but forms a shelf for the cell as a whole.

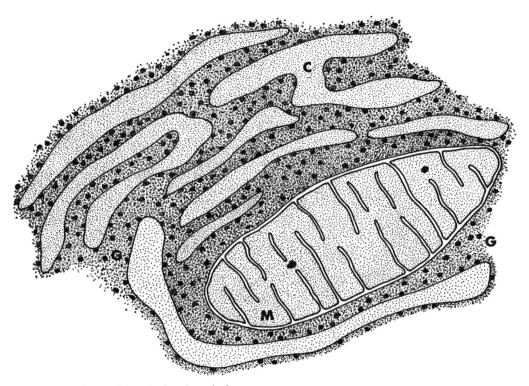

Diagram of granular endoplasmic reticulum

FIG. 6 The membrane-limited cisternae, C, have dense ribosomes located on their outer surfaces. The membrane separates the cavity from the surrounding ground substance of the cytoplasm, G. The relationship between the cisternae of the granular endoplasmic reticulum and the mitochondria, M, is often close.

In all of these examples, the diffusion distance between the mitochondria and the site of intracellular energy consumption is very small.

There are still many problems presented by the mitochondria. It appears, for example, that mitochondria contain DNA, previously thought to be located only in the nucleus, although its function is not clear. There is no doubt however that the mitochondria are the source of oxidative energy in cells of almost every type. The fine structure of the mitochondrion illustrates the uses of membranes in the cell and provides an example of the link between morphology and molecular biology.

LYSOSOMES

(Plates 10, 13, 14, 25 and 33 are relevant to this section.)

DEFINITION AND MORPHOLOGY The term *lysosome* was introduced to describe a distinctive group of subcellular particles first isolated by de Duve using the technique of cell fractionation and biochemical analysis. These particles were of the same general size as mitochondria but could be separated from the large mito-chondrial fraction of the cell by differential centrifugation. The small heavy fraction obtained in this way was shown to contain powerful hydrolytic enzymes. Since the enzyme activity was fully demonstrated only after disruptive treatment of the particles it was deduced that they were surrounded by a membrane which normally prevented leakage of their active contents. The name lysosome was intended to convey the biochemical concept of a discrete cytoplasmic particle containing hydrolytic enzymes.

The first enzyme described in the lysosome was *acid phosphatase*, but within a short period many others were found including beta galactosidase, beta glucuroni-dase, cathepsin, deoxyribonuclease and ribonuclease. The exact tally of hydrolytic enzymes varies in different tissues but a broad spectrum of activities has now been revealed in these particles. It soon became clear that the protective membrane already suggested by the biochemical behaviour of the lysosome was essential to protect the cytoplasm from its contents, since the uninhibited action of lysosomal enzymes would cause the death of the cell.

Although the lysosome can be defined in biochemical terms it has proved more difficult to provide a structural identity which can be equally clearly characterised. On electron microscopy of the particles contained in the lysosome fraction, the presence of a limiting membrane is confirmed but the contents of the lysosome show wide variations. In general, lysosomes identified biochemically correspond to structures often described in the past by electron microscopists as pleomorphic dense bodies, which are surrounded by a membrane and contain patches of dense osmiophilic material, membrane lamellar structures and pale amorphous areas. It is now possible to carry out cytochemical tests for acid phosphatase and other enzymes using electron microscopy and demonstrable acid phosphatase activity has become the main cytological criterion for the identification of lysosomes. Acid phosphatase activity has been described in a wide variety of structures from small homogeneous dense bodies to large complex structures, sometimes containing damaged but still recognisable parts of the cell. Membrane lamellae similar in structure to myelin are commonly seen and are thought to arise from the breakdown of phospholipid-containing structures. The *vesicle-containing body* or multi-vesicular body, a common cell component, is also believed to have some lysosomal activity. This structure consists of a single large vacuole surrounded by a membrane, within which are found a number of small membrane-bound vesicles.

FUNCTION Although most cells contain recognisable lysosomes, certain cells have become specialised to make and store lysosomes for use in the process of phagocytosis. The most familiar of these phagocytes are the macrophage and the polymorphonuclear leucocyte. In these and other cells, the probable site of lysosome synthesis is in the Golgi apparatus. In the liver cell, where lysosomes have been extensively studied, they are often termed peribiliary dense bodies, since they lie close to the bile canaliculi. Lysosomes may become greatly increased in number and complexity in damaged cells. In the cells of the small intestinal crypt after exposure to radiation, lysosomes which contain damaged cytoplasmic components and nuclear fragments appear after a few hours and may develop to remarkable sizes. Prominent lysosomes have been described in many other examples of unhealthy or degenerating cells. Structures of this type are often termed cyto-lysosomes.

In general, it seems that the lysosome represents a safe and convenient method of storing a number of potent destructive enzymes in the cytoplasm. These enzymes are employed in different ways, ranging from the organised activity of phagocytosis in specialised cells, where they are used to digest foreign materials, to the disposal of damaged components of the cytoplasm. The more complex forms of lysosome appear to represent the indigestible residue remaining after an episode of lysosomal activity and for this reason are often called residual bodies.

It has been suggested that the lysosomes may take part in bringing about the death of the cell. Cell death is an important part of growth and development since the remodelling of tissues and organs can take place only by the planned death and removal of unwanted cells. The death of the cell in these circumstances might result from the purposeful breakdown of the lysosomal membrane leading to release of the hydrolytic enzymes into the cytoplasm. The explosive release of lysosomal enzymes from leucocyte granules when bacteria are ingested also leads to the death of the cell. Lysosomes may be concerned in other aspects of cell behaviour. The acrosome of the spermatozoon appears to be a type of large lysosome, perhaps concerned with penetration of the ovum. It has been suggested that lysosome rupture may be a stimulus to cell division and that abnormal lysosomal behaviour could even be a factor in the production of certain forms of disease, including cancer.

CYTOPLASMIC FIBRILS AND MICROTUBULES

(*Plates* 3, 4, 6, 26, 27, 29, 34a, 38, 44a *and* 46 *are relevant to this section.*)

Cytoplasmic *fibrils* or *microfibrils* can be seen by electron microscopy in the cytoplasm of many cells. In most cases this material consists of aggregates of elongated protein molecules which are generally believed to form a diffuse skeleton extending through the cytoplasm, giving resilience and if necessary rigidity to cell structure and perhaps forming a framework for the attachment of enzyme groups and cytoplasmic structures. The common association of cytoplasmic fibrils in epithelial cells with the junctional complexes and desmosomes, forming a so-called *cell web* or fibrillar network believed to be continuous throughout the cytoplasm, lends support to the belief that such fibrils are of constructional significance, but there is little firm evidence regarding their nature. In other cases, however, cytoplasmic fibrils form an important part of cell structure and are related to function in more obvious ways. In muscle, the contracting mechanism in the cytoplasm is made up of protein fibrils with a specific chemical nature and anatomical arrangement. A fibrillar protein material isolated from amoebae may be concerned with amoeboid movement. In skin the fibrillar material which accumulates in the cell is keratin which has an important protective function. Prominent fibrils are also seen in the processes of nerve cells where they take on a pronounced longitudinal arrangement.

C

The presence of microfibrils in a cell in electron micrographs in itself is of no specific functional significance without further information concerning their nature and function. Many different proteins with different biological functions can aggregate to produce a fibrillar pattern.

With the introduction of glutaraldehyde fixation for electron microscopy, fine tubules with a diameter of around 230 Å and of indefinite length have been identified in the cytoplasmic matrix. These *microtubules*, less obvious with osmium fixation, are not composed of true membranes since they do not appear to have the characteristic trilaminar structure. On cross section a microtubule appears as a circular profile, the wall of which at high resolution may sometimes appear as a ring of some thirteen minute subunits. In longitudinal section the tubules are sufficiently narrow for their whole thickness to be included in the section and at times they can be followed for substantial distances through the cytoplasm despite their slightly irregular course.

Microtubules have been found to form the mitotic spindle, a birefringent double cone which is seen by light microscopy and radiates out from the centrioles to the chromosomes during the process of cell division. The microtubules appear to run between the centrioles and the chromosomes, suggesting that they play some part in controlling the movements of the chromosomes during metaphase and anaphase. Prominent microtubules, known usually as *neurotubules*, run parallel to the long axis of the axon or dendrite in the nerve cell. Microtubules are relatively plentiful in irregularly shaped cells such as the podocyte of the renal glomerulus. It has been suggested that the microtubule could be concerned either with the maintenance of the particular shape of the cell, in a structural role, or with the establishment of diffusion channels through the cytoplasm. Since microtubules appear also in the central cores of motile structures such as cilia and sperm tails, they may possibly be simple contractile units. This would be consistent with the apparent relationship to chromosome movement which is shown by the microtubules of the mitotic spindle. In cells in interphase it is possible that a sparse population of microtubules might simply represent residual components of the mitotic spindle remaining after the preceding cell division. The true significance and nature of the microtubule is still far from clear.

CENTRIOLES
(*Plates* 12, 13b, 44 *and* 45 *are relevant to this section.*)
MORPHOLOGY On light microscopy of many cells, an area of the cytoplasm can be recognised which appears more homogeneous than its surroundings, is free from mitochondria, and has two tiny spots at its centre. This area of the cell, often associated with the Golgi apparatus, is known as the centrosome and its two central spots are the *centrioles*. On electron microscopy, each centriole is a cylindrical structure usually around 5,000 Å in length and 1500 Å in diameter. The two centrioles normally seen in the resting cell lie commonly with their long axes at right angles to each other. The fine structural appearance of a single centriole in a thin section depends on whether the cylinder is cut in its long axis, obliquely, or

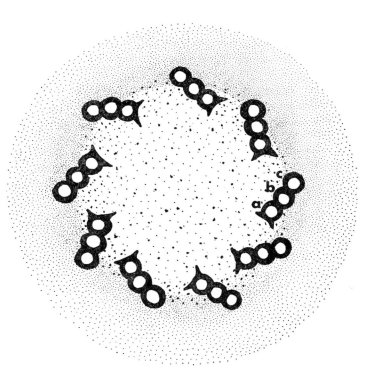

Diagram of centriolar subunits

FIG. 7 The common form of centriolar pattern, seen in cross section, consists of nine components or triplets, each composed of three adjacent tubular elements arranged in a characteristic pattern and often designated a to c as shown. Small arms may be seen stretching from subunit a towards the centre of the centriole and towards subunit c of the adjacent triplet.

transversely. In transverse section the wall of the cylinder is found to be composed of nine sets of microtubular structures which run the length of the centriole in a gentle spiral. Since transverse section shows these structures 'end on', their detailed organisation can be distinguished (Fig. 7). Each set consists of three subunits which are termed subfibrils a, b and c, arranged in a characteristic pattern. Subfibril a of each triplet has two short arms, one connecting to subfibril c of the adjacent triplet, the other reaching in to the centre of the centriole. When the cylinder is cut in its long axis, the parallel 'walls' are seen but there is little indication of their detailed structure. The centre of the cylinder may be occupied by dense material and one end often appears closed. Dense satellites which may be attached to the outside of the centrioles could serve as points of insertion for microtubules related to the mitotic spindle.

FUNCTION The centrioles are best known to cytologists on account of two aspects of their behaviour. Before cell division takes place the centrioles reproduce themselves and take up positions at opposite sides of the nucleus where they form the two poles of the mitotic spindle. In ciliated cells on the other hand the centrioles divide repeatedly and form the basal bodies which give rise to the cilia, acting as their anchorages in the cell and perhaps also as centres for their control. When centriolar division takes place, the daughter centriole grows from the parent at right angles to its long axis, perhaps explaining the consistent orientation of the two centrioles of each pair. The mode of replication of the centrioles, the significance of their internal fine structure, their association with the chromosomes at mitosis and their relation to the microtubules which comprise the spindle are still not fully understood. The association of centrioles with chromosomal movement and their relationship to cilia and flagella suggest that the centrioles could be concerned with the organisation of movement at the subcellular level.

The reason for the universal occurrence of nine subunits in centrioles is unknown. The constancy of this pattern may reflect the preservation throughout evolution of a satisfactory structural formula developed for an essential function at an early stage. A self-replicating structure which became associated with cell division in primitive forms of life might be expected to remain substantially unchanged during the subsequent course of evolution.

CHAPTER 4
Structure of Cells with Specialised Functions

The parts of the cell which have been described in the preceding chapters are the common structural units from which all cells are constructed. It is now proposed to describe the patterns of cell structure which form the 'machinery' devoted to specific tasks in specialised cells. The examples have been chosen to illustrate the relationship between fine structure and cell function in a variety of situations without an attempt being made at a comprehensive survey of the different organs and tissues of the body.

SECRETION
The process of secretion is the elaboration by a cell, from precursor materials, of a new substance which is then released from the cell. Special groups of cells gathered together for the purpose of secretion are known as glands, but although gland cells are particularly concerned with secretion, other types of cell may also show this activity. The formation of collagen by connective tissue cells is as much a secretory process as the production of digestive enzymes by the pancreas. A secretory product may be stored in the cell for some time in the form of secretory granules which often have a characteristic morphology. These are eventually discharged from the cell for use elsewhere. The fine structure of cells engaged in secretion has been found to vary in characteristic ways according to the nature of the secretion.

Glands fall into two groups, exocrine and endocrine. The exocrine gland passes its secretion through a duct on to a surface or hollow organ in the body. The salivary glands and the exocrine pancreas which secrete digestive enzymes are examples of exocrine function, but other exocrine glands can form secretions of different chemical natures. The endocrine gland secretes a chemical messenger or hormone which passes from the gland cell into the blood stream to be carried to its destination. The pituitary and adrenal are examples of endocrine glands. Hormones have different chemical compositions and produce different specific biological effects.

PROTEIN SECRETION (*Plates*, 7, 9, 22, 23, *and* 33 *are relevant to this section.*) Protein molecules perform numerous metabolic and mechanical functions in the living organism. All enzymes including the digestive enzymes released into the gut are protein molecules, so that the main function of the zymogenic cells of the gastric glands and the pancreas is the secretion of protein. The function of protein secretion is not confined to exocrine glands since the endocrine thyroid gland manufactures, for storage in the gland, a protein conjugated with its specific iodine-containing hormone. Examples of cells which secrete proteins but are not contained in glands are the fibroblast which makes the structural protein molecules of collagen and the plasma cell which produces antibody protein as part of the defences of the body against foreign materials. Protein secretion is a widespread and important cell function and the secreting cell is specifically equipped for the purpose in fine structural terms.

The essential feature of the protein-secreting cell is the presence of an elaborate system of granular endoplasmic reticulum which may often fill most of the cytoplasm. The cisternae are long, often extensively interconnected and at times so closely packed that the cytoplasmic space containing the ribosomes between adjacent cisternae is narrower than the width of the individual cisternae. The interconnection between the cisternae may allow a widespread continuity throughout the system, while the connections seen at times between the cytoplasmic cisternae and the perinuclear cisterna emphasise the dynamic unity of these components.

The contents of the cisternae vary in appearance in different cells, but are usually of low density in comparison with the adjacent cytoplasmic ground substance with its numerous ribosomes. The material contained in the cisternae may be flocculent and the cisternae dilated, suggesting that the protein synthesised by the cell can accumulate and possibly be stored within the cisternae. In certain cells, such as the pancreatic zymogenic cells of a few species, intracisternal zymogenic granules are present, but in general the material in the endoplasmic reticulum is not organised and the secretion granules originate within the Golgi apparatus. In a cell with a substantial endoplasmic reticulum the membranes of the cisternae effectively partition the cytoplasm into two phases, the interconnecting cavities of the endoplasmic reticulum, related to the inner nuclear membrane through the perinuclear cisterna, and the ground substance containing the ribosomes and other cytoplasmic components, related to the nucleus through the nuclear pores.

In cells such as the zymogenic type which secrete protein in the form of granules discharged at the cell surface, the Golgi apparatus is prominent and usually lies between the nucleus and the cell apex, where it forms a complex horseshoe pattern not always apparent in thin sections. The lamellae of the apparatus are long and the vacuoles often prominent. The Golgi apparatus is generally surrounded by a 'shell' of endoplasmic reticulum the inner cisternae of which lie close to the outer aspect of the apparatus. It is possible that transfer of newly synthesised protein material from the cavities of the cisternae to the Golgi apparatus takes place here by the production of small blebs or buds similar to micropinocytotic vesicles, which form from the membrane of the cisterna and which fill with its contents, finally pinching off to form a free vesicle (Fig. 8). The vesicle may then pass to the Golgi apparatus and fuse with its outer aspect, releasing its contents into the Golgi sacs. The components of the Golgi apparatus appear to be structurally and functionally polarised, since the formation of the secretion granules tends to occur on the opposite side of the apparatus from the endoplasmic reticulum.

The final secretion granule often seems to form by the accumulation of material of appreciable density within a Golgi sac (Fig. 8). The formed granules which lie closest to the Golgi apparatus are often paler and more mottled in texture than those which have been released from the apparatus and have passed to their storage area in the cell apex. This suggests that the granules may continue to mature, perhaps by withdrawal of fluid or by the continuing action of enzyme systems associated

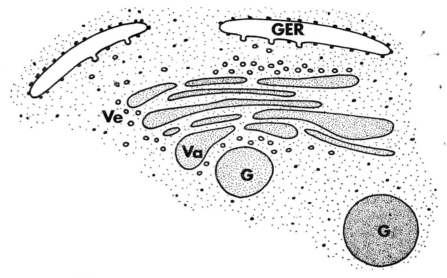

Diagram of the Golgi apparatus

FIG. 8 The lamellar pattern formed by the Golgi membranes is often prominent. Golgi vacuoles, Va, and small Golgi vesicles, Ve, are shown. Granules of secretion, G, form at the Golgi apparatus and pass to other parts of the cell. These granules may become more dense as they mature. The granular endoplasmic reticulum, GER, is often closely related to the Golgi apparatus and Golgi vesicles may form by budding from these cisternae.

with the Golgi membranes from which the membrane surrounding the granule originated. The mature secretion granules are commonly released at the cell surface by fusion of their surrounding membranes with the surface of the cell, allowing their contents to be discharged without breakdown of the structural integrity of the cell (Fig. 9). This process is known as merocrine or eccrine secretion. The dense contents of the secretion granule often undergo a change to a pale flocculent form. probably due to the uptake of water, as soon as the granule is exposed to the extra-cellular fluid. The apical surface of the protein-secreting gland cell often displays sparse microvilli, the significance of which is not clear, since they do not seem to be associated with absorption.

The most extensively studied protein-secreting cell is the zymogenic cell of the exocrine pancreas, which is similar to the chief cell of the gastric gland in its fine structure. The Paneth cell of the intestinal crypt has all the morphological charac-teristics of a protein-secreting cell although the precise function of its secretion remains uncertain. The cells of the seminal vesicle have similar ultrastructural features, as well as characteristic large secretion granules. The thyroid gland cell has a prominent and often widely dilated granular endoplasmic reticulum and behaves in some respects as an exocrine cell, since initially it discharges its protein-conjugated hormone secretion into storage spaces, the thyroid follicles. The endo-plasmic reticulum of the fibroblast varies in morphology depending probably upon

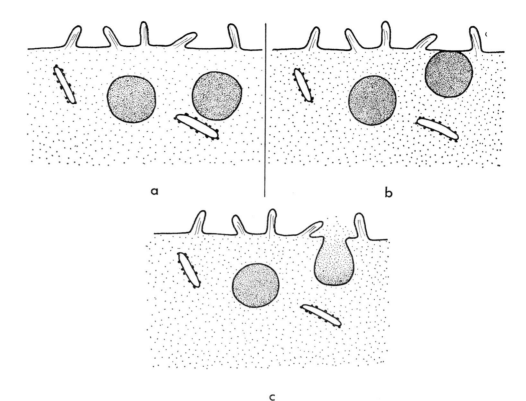

Diagram of merocrine secretion

FIG. 9 Secretion granules lying in the apex of a cell, a, may be released by fusion of the limiting membrane of the granule with the cell surface, b, and subsequent discharge of the contents of the granule, c, without any break being formed in the cell membrane.

how actively it is laying down collagen. The plasma cell, devoted to antibody protein production, shows particularly clearly the peripheral distribution of the endoplasmic reticulum surrounding the centrally placed Golgi apparatus. In the last two examples the secretory product is probably released in molecular form at the cell surface rather than as secretory granules.

ACID SECRETION (*Plate 11 is relevant to this section.*) Acid secretion is among the most remarkable of cellular functions since the physical nature of the secretion is so far removed from that of the normal body constituents. The parietal cells of the mammalian gastric gland secrete hydrochloric acid sufficiently concentrated to produce gastric contents at pH 1, a level of acidity which would cause damage and death to most cells. The essential function of the acid-secreting cell is the concentration of hydrogen ions (H^+). Since the pH of blood and tissue fluids is maintained at a level slightly higher than pH 7, and since pH is measured on a logarithmic scale, it appears that the parietal cell can concentrate H^+ by a factor of 10^6, a million times.

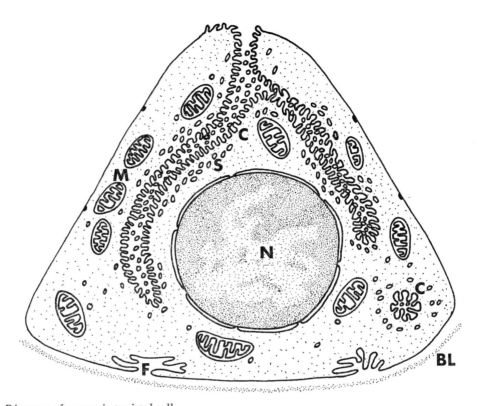

Diagram of a gastric parietal cell

FIG. 10 The main feature of the acid secreting parietal cell is the intracellular canaliculus, C, which may be seen in longitudinal or transverse section, and which may almost surround the nucleus. The many large mitochondria, M, indicate a rapid oxidative metabolism. The smooth endoplasmic reticulum, S, is apparently involved in acid secretion. Basal infoldings, F, are found, but the basal lamina, BL, does not follow the cell membrane at these points.

The gastric parietal cell has several structural adaptations related to this function, shown in diagrammatic form in Figure 10.

Surface specialisations are important in the parietal cell since the transport of ions appears to depend on the available area of the surface membrane of the cell. The intracellular canaliculus, a tubular invagination of the apical surface of the cell, is its most characteristic feature. Branches of this canaliculus ramify in the cytoplasm around the nucleus and often reach close to the base of the cell, forming an elaborate system which greatly increases the potential secretory surface of the cell and ensures that no part of the cytoplasm is far removed from the effective secretory surface of the cell. Since the lumen of the canalicular system is in continuity with the lumen of the gastric gland, acid secretion passed into any part of the system from the cytoplasm is discharged into the gland. The surface area of the canalicular system itself is increased by the presence of numerous bulbous or club-shaped microvilli which project into the canalicular lumen and may at times almost fill it with their close-packed profiles. Through these specialisations of the apical surface of the cell, an adequate area for secretory transfer is attained despite the limited apical portion of each cell which is able to reach the restricted lumen of the gastric gland. Without this canalicular specialisation it would be impossible to house the full acid-secreting potential of the stomach within the restricted space available in the gastric glands.

The base of the parietal cell presents a further specialisation which increases the available membrane surface. Basal infoldings of the cell membrane occur at several points, although the basal lamina does not follow these infoldings but remains as a platform on which the cell stands as a whole. In certain species the extent and complexity of these folds vary with the functional state of the cell, the folds becoming more elaborate during active acid secretion.

Within the cytoplasm of the acid-secreting cell there are two major specialisations of fine structure, involving the mitochondria and the smooth endoplasmic reticulum. The mitochondria are large and numerous, occupying a large proportion of the cytoplasmic volume, indicating the considerable demands which acid secretion places on the energy source of the cell. Each mitochondrion is elaborately organised, with cristae which often extend across the full width of the cell and are packed closely together. The rate of oxidative activity in the parietal cell is well above average. The smooth endoplasmic reticulum of the parietal cell is also extensively developed, although its morphology varies according to fixation. Profiles of the smooth reticulum are distributed close to the intracellular canaliculi as well as deeper in the cytoplasm between the mitochondria. Although the details of the mechanism of acid secretion are not yet known it is thought that the membranes of the smooth endoplasmic reticulum have an important part to play. The Golgi apparatus and the granular endoplasmic reticulum on the other hand, are relatively poorly developed in the parietal cell and are presumably of little importance in acid secretion.

While in the mammalian stomach acid secretion and enzyme secretion take place in different cells, the chief cell and parietal cell, certain lower species such as birds combine the two activities in one type of gland cell. In the stomach of the hen the gastric gland cell has a well developed granular endoplasmic reticulum, a prominent Golgi apparatus and typical membrane-bound secretion granules. In addition there is a well developed system of smooth endoplasmic reticulum, the mitochondria are characteristic of an acid-secreting cell and surface membrane specialisations such as basal infoldings are prominent. Replacing an intracellular canalicular system to increase the surface area available for secretion, are deep clefts between adjacent cells which allow the sides as well as the apex of the cell to be used for ion transfer. These clefts are made possible by displacement of the junctional complex from the apex towards the base of the cell. During active secretion, the cell surface including the clefts between the cells becomes covered with the typical club-shaped microvilli of the acid-secreting cell, the basal infoldings become more elaborate and there is a reduction in the extent of the smooth reticulum in the cytoplasm. These changes can be produced by administration of histamine which stimulates acid secretion.

ENDOCRINE SECRETION (*Plate 24 is relevant to this section.*) The typical endocrine gland has no lumen, no duct and discharges its secretion into the capillaries at the base of the cells. A rich blood supply is therefore an important part of an endocrine gland and capillaries come into close contact with the endocrine secretory cells. Every cell in such a gland has at least one and often two surfaces within close reach of a capillary. The capillary endothelial walls are thin and fenestrated and the vessels are separated by minimal connective tissue spaces from the surface of the cell from which the secretion is released. Despite the close relationship between cells and circulation, there are still several barriers to be crossed. After discharge from the cell the hormone must cross the epithelial basal lamina, the connective tissue space, the capillary basal lamina and the capillary endothelium, although the presence of the fenestrations commonly seen in the capillaries of endocrine glands may ease passage across this barrier. It is probable that the hormone is dissolved in the tissue fluids after its release from the cell, and makes its subsequent journey by diffusion, assisted by the concentration gradient from the cell base to the capillary lumen.

The wide range of hormone secretions produced by the different endocrine glands explains the differences of cell structure seen on close examination of different endocrine glands, but despite these differences there is a common structural link between a number of endocrine glands including the anterior pituitary, the adrenal medulla, the pancreatic islet and the enterochromaffin cells of the intestinal epithelium. In all these cells the secretory product is formed into granules in the Golgi apparatus. The small dense granules, surrounded by limiting membranes derived as usual from the Golgi membranes, accumulate in the cytoplasm. In some cases such as in the enterochromaffin or argentaffin cells of the intestine these granules show a predominant distribution towards the base of the cell and may occasionally be seen as they are discharged at the base of the cell in much the same way as individual secretion granules leave the apex of the exocrine cell.

By careful study of granule morphology the various types of endocrine cell which contain small cytoplasmic granules can be separately identified. The different cells of the anterior pituitary have distinctive patterns of granulation. The beta cells of the pancreatic islet have granules which may contain crystals with species-specific morphology. In the adrenal medulla the granules which contain adrenaline can be distinguished following glutaraldehyde fixation from these which contain noradrenaline. In certain cases the relationship of the granules to the hormone of the gland can be confirmed when secretion is stimulated, by observing degranulation of the cells accompanying a rise in the level of hormone in the blood stream. In these endocrine cells with small granules there may be few other distinctive cytoplasmic features. The mitochondria are generally small and delicate, the granular reticulum is sparse and the cytoplasm often appears relatively pale.

The secretion of steroid hormones by the cells of the adrenal cortex and by the interstitial cells of the testis is associated with several fine structural features. The mitochondria of these cells often have tubular cristae instead of the more typical shelf configuration, while a moderately well developed smooth endoplasmic reticulum is seen. Accumulations of lipid material in the form of irregular cytoplasmic inclusions are commonly encountered. The cells of the thyroid are unusual, since they combine some of the features of an exocrine as well as an endocrine gland. The initial product of the cell is thyroglobulin, a protein conjugated with the thyroid hormone. Thyroglobulin, however, is not released into the circulation, but instead is stored in the thyroid follicle, a blind vesicle which is lined by thyroid gland cells. The cytoplasm of these cells contains a prominent granular endoplasmic reticulum with dilated cisternae and a large Golgi apparatus. The thyroglobulin is released into the follicle by a process morphologically identical to exocrine secretion, but the follicles are not drained by ducts and act as storage space for hormone reserves. The cell also functions in the opposite direction as a true endocrine gland. Thyroglobulin can be removed by pinocytosis from the follicle by the thyroid gland cells, which then detach the hormone component and discharge it from the base of the cell into adjacent capillaries. The final secretory activity of the gland at any time is the resultant of these two opposing aspects of cell function.

Thus endocrine secretion is carried out by many different cells with widely differing products although common structural features can often be recognised with the electron microscope. The development of methods to identify hormones by their chemical nature using electron microscopic techniques would be of great value, since there are many aspects of endocrine function which cannot yet be fully investigated on account of the limitations of the present methods for structural study.

ABSORPTION

Small Intestine (*Plates* 4a, 5 *and* 25 *are relevant to this section.*) It is known that foodstuffs are broken down by the action of digestive enzymes secreted into the intestinal lumen. The columnar cells of the intestinal villus have been thought in the past to absorb the simple molecules resulting from this entirely extracellular, intraluminal hydrolysis. The structural specialisations of the intestinal cell were therefore interpreted as being purely absorptive in nature. It has, however, been recently confirmed that many of the enzymes thought to conduct hydrolysis in the lumen and supposed to originate from the crypt cells as the 'succus entericus' are not present in the lumen in concentrations sufficient to account for the rapid disappearance of their substrates during the normal course of digestion and absorption. It is now believed that these enzymes, such as the disaccharidases and peptidases, are in fact located in or on the surface of the cells, appearing free in the lumen only as a result of the normal physiological desquamation of cells from the villus. In this case we can now look on the fine structural specialisation of the intestinal cell as being both absorptive and digestive in nature, with digestive hydrolysis taking place at the surface of the cell as a necessary preliminary to the absorption of the products of this hydrolysis.

By light microscopy the distinctive feature of the intestinal epithelial cell is the presence of the *striated border*, a refractile zone between 1 and 2 μ thick at the apical surface of the cell. In view of its critical position forming the interface between the cells and the contents of the lumen, the striated border was extensively studied by light microscopy, but its true nature could not be firmly established, since the details of its structure were beyond the limit of resolution. The electron microscope resolved this controversy by showing that the surface of the intestinal cell is covered by a highly organised array of parallel finger-like projections which were named *microvilli*, extending into the lumen from the surface of the cell. Each microvillus measures about 1 μ in length and 0·1 μ in diameter and consists of a cytoplasmic core with a surface formed by the surface membrane of the cell. As many as 1700 of these processes may be present on the surface of a single cell increasing the area of its apical surface more than twentyfold. The microvilli are longer and more numerous towards the tip of the intestinal villus where absorption is most rapid.

When the microvillus is seen in longitudinal section a central core of longitudinal fibrils or narrow tubules from 30 to 60 Å in diameter can be seen extending from the tip of the microvillus to the cytoplasm at its root. On cross section, central fibrils are found randomly spaced within the core, about 40 in number. This is quite unlike the organised pattern seen in cilia which are much larger and more complex structures with a clearly defined constant internal pattern. These features allow cilia to be clearly distinguished from microvilli. The cores of the microvilli are linked together at their bases by a transverse network of felted fibrils extending across the cell apex between the junctional complexes, forming a region of the cytoplasm, called the *terminal web*, which is free from formed structures. The microvillous border and the terminal web together form a structural unit which can be isolated by fractionation of intestinal cell homogenates. The fibrillar network of the terminal web gives rigidity to the apex of the cell and prevents deformation of the outlines both of the individual cell and of the epithelial surface as a whole, in this otherwise delicate mucosa.

Perhaps the most significant part of the microvillous border is the apical surface membrane of the intestinal cell through which all absorbed material must pass, assuming that the junctional complex maintains adhesion between cells, presenting an effective barrier. The membrane covering the microvilli shows the typical trilaminar structure but is about 105 Å in thickness, markedly thicker than most other biological membranes such as those of the cytoplasmic organelles. The membrane at the apex of the intestinal cell is thicker than the membrane which forms its lateral and basal surface. Biochemical investigation suggests that the membrane covering the microvilli has a mosaic of enzymes built into its structure, perhaps in the form of structural assemblies. The presence of subunits of around 70 Å in diameter is now suspected from negative staining studies of isolated microvilli. The increased membrane surface provided by the microvilli at the apex of the intestinal cell can perhaps be seen as a specialisation which makes available a greater surface area to accommodate larger numbers of digestive enzyme assemblies at the absorptive surface between cell and lumen. Some enzymes might be located partly in the mucopolysaccharide fuzzy coat which covers the microvilli.

The intestinal cell has a moderately well developed smooth endoplasmic reticulum with strands of rather poorly organised granular reticulum and some free ribosomes. The Golgi apparatus lies just above the nucleus forming a complex system in which dilated Golgi vacuoles are often prominent. The mitochondria have no special features, being elongated in outline with variable internal organisation. The nucleus is basally placed, of moderate size and oval outline with a diffuse chromatin pattern. As is typical of epithelial cells, the intestinal cells are joined at their apices by junctional complexes and along their contact surfaces by intermittent adhesion plaques or desmosomes.

The intestinal epithelial cells show variations of fine structure which may depend partly on the functional state of the cell and partly on its maturity, as determined by its position on the intestinal villus. The cells of the villus are replaced constantly by division of precursor cells in the crypts and migration of these newly-formed cells to the tip of the villus, where extrusion occurs after a 'working life' of only two or three days. Thus the cells at the tip are more mature than those in the crypt and at the base of the villus. The precursor cells of the crypt have smaller microvilli, less closely packed than those of the mature absorptive cells. The crypt cell cytoplasm shows numerous free ribosomes and little formed endoplasmic reticulum, characteristic features of a 'progenitor' cell type. As the crypt cell passes on to the villus it begins to acquire the features of the mature cell while losing the potential for continued division. This structural maturation is accompanied by a development of enzyme systems and an increase in absorptive capacity.

The absorption of fat has been studied by electron microscopy, since fat droplets, when fixed, become clearly visible. Early work suggested that fat droplets could be taken up directly from the lumen in micropinocytotic vesicles which formed at the roots of microvilli, but it is now thought that this pathway is not significant in quantitative terms. Fat absorption occurs to a greater extent in the form of lipid micelles less than 100 Å in diameter held in suspension by bile salts and fatty acids in the intestinal lumen. The lipid is probably broken down at the surface of the cell and resynthesised in the endoplasmic reticulum which contains the necessary enzymes. Lipid droplets then appear in the cisternae of the endoplasmic reticulum through which they are transported in the cell and can pass to the Golgi apparatus. Chylomicrons are made in the intestinal cell by the production of a lipoprotein envelope for the lipid droplets, which then pass out of the cell to the lacteal in the core of the villus. Although pinocytosis is now not thought to be particularly important in fat absorption in adults, it does seem to form a pathway for the uptake of antibodies by young mammals. The antibodies ingested in the maternal milk are transferred unaltered to the infant circulation and give passive immunity to disease.

Abnormalities of the highly specialised intestinal epithelial cell may cause impairment of absorption of foodstuffs. If a single enzyme such as lactase is absent from the surface membrane of the microvilli, quite a common abnormality, there is failure to digest and absorb lactose from the diet. This can lead to malnutrition and diarrhoea in infancy, but the absence of the single enzyme may cause little significant ultrastructural change in the epithelial cells. On the other hand, certain individuals are sensitive to wheat protein, gluten, which damages the intestinal epithelium causing severe histological and fine structural abnormalities such as disorganisation and shortening of the microvilli. This type of abnormality, called idiopathic steatorrhoea in adults and coeliac disease when it occurs in children, is generally termed gluten-enteropathy. Gluten-enteropathy is characterised by the presence of a malabsorption syndrome, in which all of the digestive and absorptive activities of the intestine are reduced.

KIDNEY (*Plates* 39, 40 *and* 41a *are relevant to this section.*) In the renal tubule many different aspects of absorption and secretion are taking place simultaneously during the formation of the urine. Fine structural differences reflect functional variations between the cells in different segments of the nephron. The formation of urine begins in the glomerulus with the filtration of the plasma, resulting in the production of a dilute urine which collects in the urinary space of Bowman's capsule and passes into the first part of the tubular nephron. In man, 120 c.c. of glomerular filtrate are formed per minute, of which 99 per cent is reabsorbed, along with many of the solutes it contains, by the renal tubular epithelium. Most of this reabsorption takes place in the proximal convoluted tubule, the cells of which are highly specialised in fine structural terms.

The epithelium lining the proximal convoluted tubule is columnar or cuboidal in shape and the cells have a striated border on light microscopic examination. As in the intestinal epithelium the striated border consists of numerous closely packed microvilli, each one too thin to be observed individually by light microscopy. These microvilli are embedded in a dense surrounding layer reminiscent of the apical surface coat of the intestinal cell. Tubular invaginations of the cell surface between the microvilli are common in this type of cell. These have been shown to provide channels for the uptake of electron-dense materials from the nephron, such as ferritin and haemoglobin aggregates, suggesting that pinocytotic absorption of materials may be possible at this site. The morphology of these apical channels can be affected by osmotic variations.

The transport of water and ions across the cell, so important a part of the function of the proximal convoluted tubule, is reflected in the basal membrane specialisations which are prominent. The cell base is infolded to a remarkable degree, the invaginated paired membranes extending from the cell base deeply into the cytoplasm. Marked ATP-ase activity has been shown to exist at the cell base, suggesting the presence in the membrane of an energy-consuming active transport system. The large and numerous mitochondria of these cells are packed closely between the infolded membranes, with an orientation predominantly parallel to the invaginations of the cell base (Fig. 5). Elaborate infoldings of the cell base in association with fluid or ion transport are also seen in the gastric parietal cell, the striated ducts of salivary glands and the salt gland of seabirds.

PHAGOCYTOSIS

(*Plates* 13a, 13b *and* 14 *are relevant to this section.*)

Certain cells have the ability to pick up particulate material which becomes segregated in their cytoplasm in specialised structures designed for intracellular digestion. The process of uptake is termed *phagocytosis*. In theory, pinocytosis refers to the uptake of fluids while phagocytosis implies particulate uptake, but this distinction often appears artificial. The amoeba feeds itself by phagocytosis, but in the more complex animals phagocytosis has become the special function of a limited number of cell types. The most important of these are the leucocytes or white blood cells and the macrophages, which are widely encountered throughout different organs and form part of the reticulo-endothelial system of the body. The wide distribution of macrophages can be demonstrated by injecting into an experimental animal a solution of carbon particles, dyestuff, or colloidal heavy metal such as gold or thorium. Subsequent examination by light or electron microscopy shows the presence of the particles within cells with phagocytic powers. Phagocytes ingest bacteria and debris from damaged cells as well as a wide variety of other foreign materials. The sequel to phagocytosis is the enzymatic hydrolysis or digestion within the cell of the ingested material.

The most important cytoplasmic component of the macrophage is the lysosome. These inclusions contain the powerful hydrolytic enzymes responsible for the controlled digestion of particles taken up by phagocytes. In a sense the lysosomes can be regarded as the essential product of the phagocyte, a form of secretion which is not discharged, but performs its actions within the cell. The *primary lysosomes* are newly-formed simple membrane-bound structures with homogeneous or finely granular contents. As with many other cell products, primary lysosomes are probably produced in their final form by the Golgi apparatus. Apart from lysosomes the macrophage has few other distinctive cytoplasmic features. When the macrophage is stimulated by materials which induce phagocytic activity, such as experimentally injected colloidal solutions, the surface of the cell when observed by light microscopy becomes ruffled and irregular, throwing out surface projections which cause the outlines of the cell to fluctuate. Electron microscopy shows irregular indentations and projections of the cell surface membrane. The injected colloidal material adheres to the cell surface which then becomes invaginated to form a cytoplasmic vesicle, lined by the patch of membrane which was originally exposed. In this way the macrophage, by phagocytosis, ingests extracellular materials. A mucopolysaccharide cell coat can be demonstrated by special staining methods. The cell coat might regulate the phagocytic activity of the macrophage.

Digestion follows phagocytic uptake. The material ingested by the macrophage is brought into contact with the enzymes of a lysosome. If the material is in the form of small particles they may be directly incorporated within a lysosome. With larger particles, such as bacteria engulfed by leucocytes, the lysosomes are discharged into the cytoplasmic vacuole which encloses the material, thus forming a kind of intracellular stomach, full of active digestive enzymes but isolated from the cytoplasm by the persisting membrane of the vacuole. This activity generally causes the death of a polymorph, but in the macrophage digestion can normally proceed without danger to the components of the cytoplasm. The products of hydrolysis can finally be absorbed into the cytoplasm of the macrophage while any indigestible residue accumulates within the lysosomal structure. The complex bodies which are formed by the combination of ingested material with lysosomal enzymes during active phagocytosis are termed *secondary lysosomes*, while the structures containing undigested residues are known as residual bodies. The wide range of materials ingested from time to time by phagocytic cells leads to a wide range of variation in the strucure of the residual bodies. Dense granular areas and membrane lamellae are common components and they are often pigmented, giving rise to the appearance of granular deposits with a brown colouration under light microscopy.

The macrophages of the reticulo-endothelial system form a network throughout the body which reacts in a specialised way to the presence of foreign or unwanted material, accomplishing as far as possible its removal and destruction. Phagocytes act as scavengers, removing debris when cells are injured by disease or die a natural death and clearing the circulation of foreign materials which may enter the body. Phagocytosis of bacteria by white blood cells forms an important part of the defences of the body against infection.

PERMEABILITY

In certain situations the rapid passage either of nutrient materials or of waste products of metabolism across cellular boundaries is of great importance. One of the main constants governing the rate of diffusion across a barrier is the thickness and permeability of the barrier. In the capillaries, the lungs and the renal glomerulus, where materials must cross cell barriers, the block to diffusion is reduced to a minimum by thinning of the cells which form the barrier. These cells, specialised to allow permeability, will now be examined.

CAPILLARIES (*Plates* 17, 18, 21a and 36 *are relevant to this section.*) The blood capillaries, the smallest vessels of the circulation, form the main barrier throughout the body between the circulating blood and the cells of the various tissues. The oxygen and essential metabolites without which the cell cannot survive must pass across the capillary wall, while waste products of metabolism must be removed from the tissues. The capillary (Fig. 11), is a delicate tubule composed of flattened endothelial cells joined together by adhesion specialisations along their contact or meeting edges. Each cell has a thickened area which contains the nucleus along with a few of the common cytoplasmic components such as a small packet of Golgi membranes, a few vesicles and occasional mitochondria. The capillary endothelial cell is surrounded by a closely applied basal lamina, around which are arranged connective tissue elements with collagen fibres tending to fuse with the basal lamina. There is therefore a connective tissue space around the capillary which separates it from the cells of the organ through which it passes.

In the thinnest parts of the capillary wall, in places only a few hundred Ångstroms thick, there are two specialisations which may aid the passage of materials across the endothelial layer. The first of these specialisations, thought to represent

D

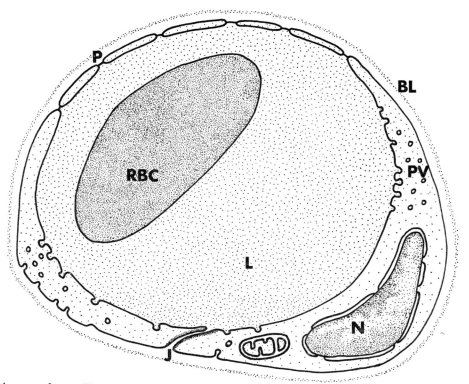

Diagram of a capillary

FIG. 11 The capillary lumen, L, is shown, containing a red blood corpuscle, RBC. The endothelial cell forming the capillary wall is very thin in places and endothelial pores, P, may be found in certain sites. Micropinocytotic vesicles, PV, may also be present. Endothelial cell junctions, J, are seen.

micropinocytotic vesicles, are flask-shaped invaginations of the inner and outer surface of the endothelial cell, known as caveolae. These vesicles, containing small quantities of extracellular or intravascular fluid may pinch off from the cell surface and form isolated spherical vesicles within the endothelial cytoplasm. These vesicles are thought to pass across the cell and fuse once again with the opposite surface membrane, discharging their contents on the other side (Fig. 2). In this way the endothelial cell could actively assist transport across the capillary wall. Thick-walled vesicles suggestive of selective uptake, as well as thin-walled caveolae are sometimes seen. The second specialisation of capillary endothelium seen in some situations and absent in others is the presence of capillary pores or fenestrations which appear as partial discontinuities several hundred Ångstroms in diameter in the endothelial lining. These pores are bridged by a diffuse diaphragm which appears thinner than the surface membrane of the endothelial cell, but still presents a significant structural barrier. Pores may appear at frequent intervals in the capillary wall.

Although it is assumed that the caveolae and endothelial pores are of importance in relation to vascular permeability, it is not clear to what extent they give either active or passive assistance to the passage of different substances. The normal vascular permeability may be due mainly to a physiological leakage at the points of junction between endothelial cells. It has been shown that small molecules can pass between the cells, suggesting that their contact specialisations are not designed to seal off completely the lumen of the vessel. During inflammation, when there is increased vascular permeability, this leakage between endothelial cells seems to be increased.

Capillary endothelium varies in structure in different sites. In the endocrine glands and the intestine where capillaries are receiving hormone secretions or absorbed food materials in addition to allowing normal metabolic interchange, fenestrations are commonly present. In the brain on the other hand, capillary endothelial fenestrations are absent and there are perivascular foot processes of neuroglial cells closely applied to the outside of the vessel. These features may provide a basis for the physiological blood-brain barrier. There may well be other differences in capillaries related to specific endothelial functions.

The capillary wall is not the only significant barrier to diffusion between the tissues and the blood. Between the circulation and the cytoplasm of an epithelial cell lie not only the capillary endothelial cell membrane and cytoplasm, but also the endothelial basal lamina, the connective tissue ground substance, the epithelial basal lamina and the membrane of the epithelial cell itself. Diffusion could be influenced at any of these points by selective processes. The full function of the capillary endothelium with its caveolae and fenestrations remains to be clearly defined.

LUNG (*Plates* 19, 20 *are relevant to this section.*) The lung is designed to provide an extensive interface between the air contained within the alveoli and the blood in the complex pulmonary capillary bed. Across this interface in opposite direction pass oxygen, essential for the metabolism of the body, and carbon dioxide, the end-product of metabolism. The interface must be strong enough to resist leakage of plasma from vessels which are poorly supported and to withstand mechanical stress during respiration and coughing, yet thin enough to allow gas exchange to continue at a rate sufficient to meet the needs of the body. The barrier in the lung between air and blood consists of the alveolar lining epithelium, the capillary endothelium and the intervening connective tissue layer.

The pulmonary capillaries are free from fenestrations but otherwise show no distinctive features. At places the capillary wall shows the presence of flask-shaped caveolae suggestive of micropinocytosis. The endothelial cells have the usual external investment of basal lamina with a surrounding, but at times narrow, connective tissue space, in which pulmonary macrophages are numerous. The capillaries are reinforced by a delicate framework of fine collagen fibres and lie within the cavity of the common wall between adjoining alveoli. The epithelial lining of the alveoli is often so closely applied to either side of the capillary that virtually no loose connective tissue can intervene between them.

It is now established that there is a complete but tenuous epithelial cell layer continuous throughout the pulmonary air spaces. The nuclei of the main cell type, the flattened alveolar lining cells, tend to lie in the corners and angles of the alveoli leaving wide areas of alveolar wall covered by their flat and often featureless cytoplasm. Since this lining, despite its thinness, is a continuous epithelial sheet, there is a typical underlying continuous epithelial basal lamina, closely applied to the cells. Although the entire barrier between blood and air may be less than 1000 Å thick, it always consists of the same distinct physical barriers, the alveolar lining epithelium, its basal lamina, the connective tissue space, the endothelial basal lamina and the endothelial cell. The connective tissue space however, may disappear or become very thin, leading to the close apposition or fusion of the two basal laminae.

The alveolar pattern of the lung presents certain physical problems. The moist epithelial surfaces of the alveoli must be separated from each other to ensure aeration, yet the considerable surface tension forces acting at the alveolar level tend to produce adhesion between alveolar walls and collapse of the alveoli. Surface tension forces in the alveoli are apparently reduced by the secretion of a surface-acting substance by a second type of cell in the alveolar epithelium, the great alveolar cell. These cells which usually lie in the angles of the alveoli are interspersed between the alveolar lining cells and are joined to them by junctional complexes. The great alveolar cells share the same basal lamina as the lining epithelium and are clearly epithelial in type. They have a flat cuboidal shape with dense cytoplasm, lamellated secretory inclusions and characteristic short irregular surface microvilli.

RENAL GLOMERULUS (*Plates* 39, 40 *and* 41a *are relevant to this section.*) In the renal glomerulus another complex relationship between endothelial and epithelial cells is found at a site where controlled permeability is an essential function. In contrast with the lung, where retention of fluid within the capillaries is essential, filtration of the circulating plasma takes place in the glomerulus. The fluid which passes from the capillary through the filtration barrier into the urinary space and the first part of the tubular system of the nephron is a dilute form of urine which is then concentrated and altered by passage along the nephron. The renal glomerulus consists of a tuft of capillaries projecting into Bowman's capsule, the dilated blind end of the nephron. The flat cells of the parietal epithelium lining Bowman's capsule become continuous with the visceral epithelium of Bowman's capsule at the point where the capillaries are invaginated into the capsule. The visceral epithelium of Bowman's capsule is a system of cells forming an investment around the individual capillaries of the glomerular tuft. Between the visceral and the parietal epithelial cells of Bowman's capsule lies a narrow cleft, the capsular space or urinary space, which is continuous with the lumen of the nephron and drains ultimately through the ureter into the bladder.

The glomerular capillaries, as might perhaps be expected, are extensively fenestrated, the pores in the endothelial cells being bridged by a tenuous diaphragm. A prominent basal lamina surrounds the capillary. Each capillary of the glomerulus is then covered by an epithelial sheath consisting of the *podocytes*, the visceral epithelial cells of Bowman's capsule. The epithelium forms not a continuous cytoplasmic layer but a complex interdigitating covering of closely packed podocyte foot processes, or pedicels, sent out in different directions by the podocytes which lie between the capillary loops. A single podocyte may have foot processes applied to the surface of several capillaries, while its nucleus and surrounding cytoplasm lie in a cleft between the vessels. Adjacent foot processes are joined by a delicate connection, the filtration slit membrane, which is the main barrier to free passage between the processes.

Since the podocytes are epithelial in nature they lie in contact with the continuous basal lamina intervening between the foot processes and the capillary endothelium. Between the epithelial and the endothelial cells there lies in fact a single shared thick basal lamina, the filtration membrane, known by histologists as the basement membrane. Although the filtration membrane may arise by fusion of an endothelial and an epithelial component, normally two separate components are not seen. It is likely that much of the material of this thick basal lamina is normally produced by the podocyte. The basal lamina is believed to form a selective barrier of great importance in glomerular filtration.

It is possible that the podocytes may do more than form a simple mechanical filter. The existence of an enclosed subpodocytic space has been suggested on the basis of detailed study of the relationships between the podocytes and the capillaries. It seems that in many cases the glomerular filtrate, having passed through the capillary endothelium and the adjacent basal lamina and between the closely packed foot processes, may enter into a space which is still enclosed by podocyte cytoplasm. The filtrate trapped in the subpodocytic space might require to cross podocyte cytoplasm to gain access to the nephron. In this way the podocyte might influence the composition of the glomerular filtrate by selective action.

The electron microscopic study of the glomerular basal lamina and the relationship between the epithelial and the endothelial components has recently become of significance in various forms of kidney disease. The electron microscope has made possible a more rational approach to the detailed interpretation of pathological changes previously barely recognised by light microscopy. Early changes not formerly visible can now be seen in glomerular disease, including thickening and patchiness of the basal lamina and fusion of the podocyte foot processes. These disturbances of fine structure are accompanied by abnormalities of the selective filtration function of the glomerulus. In this case a lesion, presumed in the past to exist on the basis of clinical and physiological observations, but beyond the limit of resolution of the light microscope, has finally been clearly pictured by electron microscopy. In the future the electron microscope may become of value in the diagnosis of particular types of glomerular disease not readily distinguished by light microscopy.

STORAGE AND CARRIAGE

The form which specialisation of function takes in certain cells is the storage or carriage of materials for widely differing purposes. In such cells the stored material may have its own specific fine structural appearances, while the formation of the material and the metabolic activity of the cell are often reflected in the structure of the cytoplasmic organelles.

RED BLOOD CORPUSCLE (*Plates* 17, 19 *and* 20 *show red blood corpuscles.*) The major function of the blood is the carriage of oxygen to the tissues. The red corpuscles of the blood have become specialised to assist this function by accumulating within their cytoplasm large quantities of haemoglobin, which can enter into a loose and readily reversible chemical association with oxygen. In this way the oxygen-carrying capacity of the blood is greatly increased. In the mammal the red corpuscles carry this specialisation to the extreme by losing their nucleus during the late stages of development. For this reason, the circulating red corpuscle is not a true cell, and consists of little more than a membrane surrounding a concentrated solution of haemoglobin within an apparently structureless matrix.

The precursors of the red corpuscle carry in their cytoplasm the biochemical apparatus for haemoglobin synthesis. Since the haemoglobin is not intended for use outside the cell, but is simply for storage in the cytoplasm, the cytoplasmic organelles are not those typical of a protein-secreting cell such as the plasma cell. However, despite the absence of a complex granular endoplasmic reticulum, the active protein synthetic function is betrayed by the numerous free ribosomes which give the early red cell precursors their characteristic cytoplasmic basophilia. The iron component of the forming haemoglobin is reflected in the presence of ferritin molecules in the cytoplasm. There is evidence that ferritin is taken up into these cells by a form of selective micropinocytosis directly visible by electron microscopy. As the haemoglobin accumulates in the maturing cell it replaces the cytoplasmic organelles. The haemoglobin concentration in the mature red corpuscle reaches as high as 33 per cent, its high density and osmium affinity being reflected in the featureless dense appearance of the red corpuscle under the electron microscope. The specialisation of the developing red corpuscle results in the production of an efficient vehicle for oxygen carriage. In the process, however, with the loss of the nucleus, the essential cellular identity of the carrier has been sacrificed in the interests of efficiency. For this reason the red corpuscle cannot reproduce itself and has a limited life span.

SKIN (*Plates* 3, 4c *and* 6 *are relevant to this section.*) The skin is so familiar that its importance as an organ of the body is often underestimated. The protection which is provided by the intact skin surface is so essential that the loss of a significant proportion of skin cover may often prove fatal. The skin protects the body from mechanical injury and from the loss of tissue fluids and proteins from deeper tissues by virtue of its continuously renewed layer structure and the accumulation of keratin within its cells. The epidermis is a stratified squamous epithelial layer supported on an elastic, resilient but strong underlying layer of connective tissue. The adhesion between the cells of the epidermis is particularly strong, adding to the capacity of the skin to resist injury.

Much of the imperviousness and resistance to attrition shown by the skin is due to the accumulation of the structural protein keratin within the maturing epidermal cell. As the cells pass towards the surface from the basal layer where the population is renewed by repeated division of progenitor cells, they begin to synthesise and accumulate keratin within their cytoplasm in a form visible by light microscopy. The synthesis of keratin is accomplished mainly by free ribosomes in the cytoplasm of the cell rather than by elaborate cytoplasmic membrane systems. The accumulated keratin eventually replaces the other cell components and when the cell dies it forms an integral part of the protective horny surface layer of the skin which is essential for its mechanical function. The eventual death of the individual cells is the price paid for this protective structural specialisation.

The strong adhesion between the epithelial cells of the epidermis is reflected in the elaborate contact specialisations as seen by electron microscopy. The presence of 'prickles' between the epidermal cells, visible by light microscopy, was formerly interpreted as an indication of the presence of intercellular bridges carrying 'tonofibrils' from cell to cell. It is now clear that these specialised areas are formed by large desmosomes into which dense bundles of cytoplasmic fibrils are inserted. There is no cytoplasmic connection between cells. The mechanical coherence of the epithelial cells is maintained by these contact specialisations to such a degree that shearing forces must be considerable to cause serious disruption in the epithelium.

ADIPOSE TISSUE The storage of fat is important in the body economy in view of its value not only as insulation but as a fuel rich in energy. The oxidation of fat produces over twice as much energy as the oxidation of a similar weight of carbohydrate or protein. Fat is normally stored in connective tissue cells and is characterised in the electron microscope by its presence in the cytoplasm without a surrounding membrane and by its homogeneous consistency. Droplets of fixed fat are often of moderate density and are particularly liable to produce sectioning artefacts on account of their hardness. The typical fat cell contains a single large fat droplet which compresses the other cytoplasmic components to the extent that they may not be clearly recognisable even with electron microscopy. When mobilisation of fat stores is required, the lipid material is withdrawn from the cells and transported to where it can be metabolised.

Brown fat is a specialised form of fat storage tissue particularly prominent in very young mammals and in animals which hibernate. In contrast to the picture described above, the brown fat cell contains a number of smaller droplets. Brown fat cells are characterised by the presence of numerous large and well organised mitochondria indicating an unusual rate of oxidative metabolism, particularly striking when compared with the small metabolic demands of conventional adipose tissue. Brown fat cells are believed to oxidise stored fat within their own cytoplasm for the purpose of heat production, to maintain body heat in the very young, poorly insulated animal or to restore the normal body temperature of the hibernating animal prior to its re-awakening.

SUPPORT

Those cells and their products which have a mainly mechanical function in the body are known as connective tissue. In connective tissue the cells are relatively sparse while the intercellular materials, both fibres and surrounding matrix, are plentiful. The mechanical strength of connective tissue resides in the fibres produced by the connective tissue cells and in certain cases in the matrix which surrounds the cells. This is in contrast with epithelial tissues in which the cells, the essential functional units, lie close together with little extracellular material.

CONNECTIVE TISSUE FIBRES (*Collagen fibres appear in plates* 6, 14, 17, 21, 29, 34a.) Collagen, the principal fibre of connective tissue, is synthesised by the fibroblast. It appears by light microscopy as dense eosinophilic bundles of coarse fibrous material which can be resolved by the electron microscope into aggregates of finer dense fibrils each of which has a characteristic repeating pattern with a number of fine cross bandings, but a major repeat of 640 Å. Collagen is believed to be secreted by the fibroblast in the form of the tropocollagen molecule, a fibrous macromolecule with a characteristically high proportion of the amino acids glycine, hydroxyproline and proline. These molecules have small areas of increased density at points along their helical structure and appear to line up in chains which may fit together in some way with a regular overlap, giving rise to the main 640 Å collagen repeat. The line-up of matching densities in adjacent tropocollagen molecules accounts for the minor striations of the collagen fibre. Collagen is particularly resistant to both chemical and physical damage and is found in a similar molecular form in many different species. Its chemical and mechanical properties are ideal for its role as the mechanical framework of connective tissue.

It is customary to distinguish dense and easily recognisable collagen, which is eosinophilic by light microscopy, from the more delicate framework of so-called reticular fibres which form around tissue elements, a network seen only by light microscopy following silver staining methods for 'reticulin'. On electron microscopy however, the individual fibres of collagen and reticulin are apparently identical and it is probable that they differ only in the size of the aggregates of fibres which are formed. Elastic fibres appear thick and homogeneous in the electron microscope in the coarse elastic tissues in which they are most easily recognised. They have irregular margins and show no evidence of periodic structure, as might be expected in material which can undergo severe distortion without suffering structural damage. Elastic fibres may be of low contrast in otherwise osmiophilic dense connective tissue, so that they stand out against the surrounding collagen.

CONNECTIVE TISSUE MATRIX Although the mucopolysaccharide matrix of connective tissue is not always distinctive on electron microscopic examination, its functions in the control of diffusion may be significant. In loose connective tissue the matrix has no organised fine structure but in bone and cartilage where the matrix has taken on a rigid mechanical function, it becomes more prominent on account of its increased density. In bone the crystals of calcium salts which are laid down in sheaves in the matrix have a distinctive electron-dense needle-shaped appearance. Collagen fibres in bone form the framework on which the dense components of the matrix are deposited and the repeating collagen pattern can still be seen in places.

CONNECTIVE TISSUE CELLS (*Plates* 1, 12, 13, 14 *and* 21a *show connective tissue cells.*) The fibroblast, the principal connective tissue cell, is the source of collagen and its prominent granular endoplasmic reticulum is in keeping with this protein-secretory function. The outlines of the cell are irregular in comparison with epithelial cells and the fibroblasts in connective tissue, in contrast with epithelial cells, do not associate with each other by forming desmosomes. Connective tissue cells tend to be scattered, each independent from the others, producing collagen locally. Among the other cells seen in connective tissue, especially in sites such as the core of the intestinal villus, are the plasma cell, eosinophil, mast cell and macrophage. The plasma cell is rounded, contains a dense nucleus with a characteristic dense clumped chromatin pattern and has an extremely elaborate granular endoplasmic reticulum and Golgi apparatus. It produces the antibody globulin molecules which form an important part of the immunological defences of the body. The eosinophil leucocyte has characteristic elliptical dense cytoplasmic granules each of which contains a 'crystalline' structure. These granules have many of the characteristics of lysosomes when they are studied biochemically. The mast cell cytoplasm is full of dense granules without specific fine structural characteristics, while the macrophage contains lysosomes of differing morphology. All of these cell types, along with other poorly defined cells classed simply as 'connective tissue cells' are again separated from each other by collagen fibres and connective tissue matrix and have no pronounced tendency to form closely related groups.

MOVEMENT
CILIA (*Plates* 5, 12, 13b, 25, 43 *and* 44 *are relevant to this section.*) While complex animals rely on muscles for movement a more delicate apparatus exists in simple unicellular creatures and elsewhere. Certain protozoa such as paramecium depend for motility upon the repeated whiplash activity of numerous thin motile projections or *cilia* which can be seen under the light microscope. The combined action of the cilia carries the cell through the surrounding fluid. In higher forms of life cilia are present in special sites where their action can be of particular value. The epithelial cells of the upper respiratory tract are ciliated, providing the surface of the trachea, bronchi and nasal passages with a continuous carpet of cilia in constant motion. Mucus secreted by goblet cells forms a surface layer which traps inhaled particles

of dust and bacteria. The action of cilia then sweeps the mucus back to the larynx where it can be removed by coughing.

Each cilium is an elongated cytoplasmic extension covered by the apical surface membrane, which measures from 5 to 10 μ in length and about 0·25 μ in diameter and is barely able to be individually resolved by light microscopy. The cilium has a central core of parallel subunits which extend, unbranching, from its base to its tip. This core is termed the *axial filament complex*, although at moderate resolution the subunits are found to be tubular in structure. The detailed arrangement of the components of the axial filament complex, obscured in longitudinal section, is clearly seen in transverse section, which allows an 'end on' view of the subunits

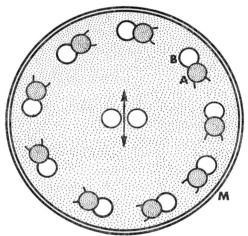

Diagram of a cilium in cross section

FIG. 12 The trilaminar structure of the cell membrane, M, which surrounds the cilium, is shown. There are nine subunits arranged in a circle, each consisting of two apparently tubular components, often designated A and B as shown. Subunit A may have small arms and is often more dense than subunit B. A central pair of tubules is always present. The direction of ciliary beat is indicated by the arrows, which lie perpendicular to a line through the centres of the two central tubules.

(Fig. 12). Two tubules, each 240 Å in diameter, lie slightly separated in the centre of the complex. Nine evenly spaced pairs of tubules are grouped peripherally around the central pair. One member of each pair, termed subfibril A, has a denser core than the other, subfibril B. Two short arms extend often from subfibril A of one doublet towards subfibril B of the adjacent pair. The components of the axial filament complex appear to originate from the *basal body*, a cylindrical structure resembling a centriole which anchors the cilium in the apex of the cell and may control and co-ordinate ciliary activity. The basal bodies of the cell arise during development from repeated division of the centrioles, explaining the fine structural similarities between centrioles and cilia.

The cilia of a single cell or of an entire region beat together with co-ordinated function, the direction of beat being defined by the orientation of the central pair of the axial filament complex. When a number of cilia are cut in cross section in a particular specimen, this orientation is approximately parallel and the direction of beat lies at right angles to it. The beat of a cilium consists of a fast forward stroke and a slow recovery stroke. The mechanism of ciliary action is uncertain, but it seems likely that the nine peripheral components bring about movement by contracting in a predetermined sequence. The two central tubules have been proposed as a conducting mechanism for an activating impulse.

Cilia and microvilli are quite different specialisations which may at times coexist on the same cell. Microvilli are fixed specialisations which increase surface area, while cilia are motile cell processes. Microvilli, too small to be individually resolved by light microscopy, form a striated border if closely packed. Cilia are large enough to be resolved as individual structures. While microvilli have a poorly organised core, the axial filament complexes of cilia are quite distinctive on electron microscopy. It is important to distinguish clearly between these different specialisations of the cell surface.

FLAGELLA AND SPERMATOZOA (*Plates* 10, 43, 44, 45 *and* 46 *are relevant to this section.*) The flagellum, a motile cell specialisation closely related to the cilium, is distinguished by its length, which can be up to 150 μ, and its spiral action. Flagella are often single and act as propulsion for certain motile cells such as the spermatozoon. In cross section the cilium and the flagellum are apparently identical. The typical thick flagella of higher species must be distinguished from bacterial flagella which are much thinner, having no organised axial filament complex.

The flagellum provides propulsion for the spermatozoon. The function of this cell is the carriage over considerable distances of the genetic information contained in the nucleus. The movement of the sperm carries it towards the ovum with which it fuses, the combined nuclear information of sperm and ovum forming the sum of the genetic composition of the new individual thus produced. The sperm has two essential parts concerned with these functions, the nucleus which carries genetic information and the propulsion unit which carries it to its destination. Since streamlining of structure is important, it is not surprising to find specialisations in the sperm designed to minimise the load which has to be carried. During its development from precursor cells in the seminiferous tubule to its mature streamlined form, the spermatozoon sheds almost all of its cytoplasm, retaining only those components which are essential to motility. The appearance of the sperm is in contrast to that of the sedentary ovum, which has a considerable cytoplasmic accumulation of material available for the cells of the early embryo.

The head of the spermatozoon contains the nucleus in which the chromatin is highly condensed, apparently existing predominantly in an inactive form. The sperm head may well represent the smallest volume in which its genetic information can be conveniently stored. The restriction of nuclear size is clearly in the interests of efficiency since it reduces drag. During the development of the sperm in the seminiferous tubule of the testis, a collection of material produced by the Golgi apparatus accumulates in the acrosome, a cap which becomes applied to the pole of the nucleus. The acrosome appears to be a specialised lysosome, which may be concerned in penetrating the ovum at fertilisation and initiating division. Apart from these structures contained in the sperm head, all the remaining components are concerned with motility.

The tail of the spermatozoon contains the axial filament complex, identical in cross section to that of cilia and flagella in other sites. The axial filaments or tubules arise from a centriole at the posterior pole of the nucleus and run to the tip of the tail. In the mammalian sperm nine additional broad dense components, clearly seen on cross section, are arranged in a ring external to the nine tubular doublets of the axial filament complex. These additional components, straight and unbranching on longitudinal section, terminate at different levels along the tail of the sperm, giving rise to different appearances of the tail on cross section at different levels along its length. They are asymmetrical in their arrangement, and can be numbered according to a simple convention. It is believed that these additional components of the tail might be extra contractile elements reinforcing the axial complex.

Movement can be accomplished only by the expenditure of energy. The demand for energy to sustain movement must be met from mitochondria within the cell by ATP produced by oxidative phosphorylation. The midpiece of the sperm consists of a tightly-wound spiral of mitochondria which forms a coil through the core of which pass the components of the axial filament complex. The raw material for oxidation is taken up from the surrounding environment and broken down by the mitochondrial enzymes to generate energy in the form of high energy phosphate bonds, readily available for conversion into mechanical work. Further along the tail, the mitochondrial sheath is replaced and the cell perhaps strengthened by a system of fibrous ribs linked longitudinally in the plane of the central filaments.

CONTRACTION

One of the basic functions of protoplasm is contraction, for which the muscles of multicellular animals have become specialised. Muscle is the source of movement, either for locomotion, by its action on the bones of the skeleton, or for internal motility, by its control over the diameter of blood vessels and hollow organs. Three main groups of muscular tissues are recognised by their histological characteristics, skeletal or striated muscle attached principally to the bones of the skeleton, visceral or smooth muscle found principally in the viscera or internal organs and cardiac muscle which is present exclusively in the heart.

The cellular unit in skeletal muscle is the individual muscle cell which is often referred to as a muscle 'fibre' on account of its elongated shape. The term fibre, however, is perhaps best reserved for extracellular structures with a filamentous pattern such as collagen. A single skeletal muscle cell may be centimetres long, extending at times from end to end of the muscle, and may be up to 100μ thick. Along the length of the cell there are numerous peripherally placed nuclei which lie close to the cell surface. A regular pattern of cross striation with a spacing of 2 to 3μ is the most prominent feature of the skeletal muscle cell, and is the basis of the term 'striated muscle' commonly applied to muscle cells of this type. Skeletal or striated muscle is under the control of the cranial or spinal motor nerves and is dependent on the integrity of its innervation for its continued function.

The unit of visceral muscle is the muscle cell which in this case has a single nucleus lying in its centre. The cells of visceral muscle have a wide range of size from 10 to 100 μ in length and up to 10 μ thick, the precise dimensions depending on the location and function of the cell. Visceral muscle shows no cross-striated pattern and for this reason is often described as smooth muscle. Visceral or smooth muscle is innervated by autonomic nerves and in general is not under voluntary control. The cells have an inherent rhythmic activity not entirely dependent on the integrity of external innervation and are responsive also to stimulation by certain hormones.

Cardiac muscle has a branching structure with centrally placed nuclei. The nuclei of neighbouring cells are separated by partitions known as intercalated discs. The cells are cross-striated with a sarcomere pattern similar to that of skeletal muscle. Cardiac muscle, however, has an inherent rhythmic activity, modified by the action of nerves but not dependent on them for its function.

SKELETAL MUSCLE (*Plates 26, 27, 28 and 32, are relevant to this section.*) Several components are recognised by light microscopy in the skeletal muscle cell. Each cell contains a large number of longitudinally arranged fine parallel *myofibrils*, in the order of 1 μ in diameter, packed together so that they are superimposed on each other in histological sections and appear as a faint longitudinal striation. Each fibril is individually cross-striated, the cross striation of the muscle cell as a whole resulting from the alignment of the striations of individual parallel myofibrils. A small amount of cytoplasmic material known as *sarcoplasm* surrounds the myofibrils and separates them from each other. Mitochondria can be detected in the sarcoplasm by light microscopy with special stains. The nuclei lie within the sarcoplasm in close contact with the outermost myofibrils. A delicate investment which appears to surround the cell is referred to as the sarcolemma.

In fine structural terms the muscle cell, although specialised in many ways, conforms to the basic rules of cellular construction. A single typical cell membrane invests the cell forming an unbroken partition between the cytoplasm and the tissue fluid. At points along the surface of the muscle cell small flask-shaped invaginations of the membrane suggestive of micropinocytosis may be observed. Immediately external to the membrane a distinct cell coat or basal lamina is closely applied to each individual cell, surrounding it and separating it from its neighbours, reinforced in places by a delicate lacework of collagen fibres. This combination of structural components forms the image interpreted in light microscopy as the sarcolemma. Within the cell the parallel myofibrils are surrounded by organised sarcoplasm.

Within the individual myofibril which is perhaps 1 μ in diameter but extends the length of the cell, an organised longitudinal system of parallel thick and thin filaments can be seen by electron microscopy (Fig. 13). The alignment of these filaments forming regular regions of overlap is responsible for the distinctive repeating pattern of cross striations and is believed to form the molecular basis for muscle contraction. A complex nomenclature exists to identify the components of the *sarcomere*, the repeating unit of cross striation. The region of the myofibril in which only thin filaments are found is the I band and the region in which thick

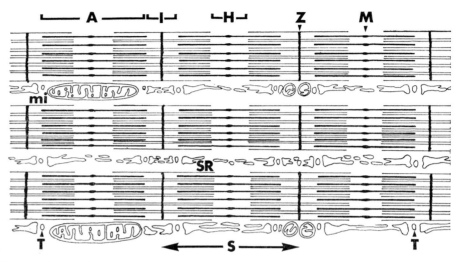

Diagram of striated muscle

FIG. 13 Three parallel myofibrils are shown separated by sarcoplasm which contains the sarcoplasmic reticulum, SR, with associated T tubules, T, and mitochondria, mi. S indicates the length of a sarcomere, extending from one Z line to the next. The thick filaments and thin filaments overlap to form the different bands of the repeating pattern of muscle, A, H and I. The M and Z lines are indicated. This diagram indicating the resting state should be compared with Figure 14, which shows a myofibril in contraction.

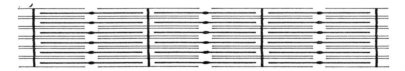

Diagram of a contracted myofibril

FIG. 14 Although the thick and thin filaments remain the same length as in the resting state shown in Figure 13, their area of overlap is increased causing shortening of the I band and H zone. However, the A band remains the same length. The sarcomere length is thus reduced and the muscle as a whole is shortened.

filaments are present is the A band. The myofibril is composed of alternating I bands and A bands. The I band is bisected by the dense Z line which links together the mid points of the thin filaments. The sarcomere extends from one Z line to the next and consists therefore of two half I bands with a complete A band between them. In the A band of the relaxed myofibril the central portion, the H zone, contains only thick filaments. Thickenings at the mid point of each thick filament align to form the M line which bisects the A band. At each end of the A band there is overlap between the interdigitating thick and thin filaments. At intervals there appear to be minute cross-linkages which join overlapping thick and thin filaments together.

When a transverse section of muscle is examined rather than a longitudinal section, the filaments of the myofibril are viewed 'end on' rather than 'side on' and appear in cross section as dense spots. The thin filaments are around 60 Å in diameter, the thick filaments about 120 Å. If the plane of section passes through the I band where only thin filaments are present, a pattern of spots each 60 Å in diameter is seen. If the section cuts the central part of the A band where only thick filaments are found, cross sections of thick filaments each 120 Å in diameter are seen. If the section passes instead through the region of overlapping thick and thin filaments, a hexagonal pattern can often be seen in which each thick filament is surrounded by six thin filaments. In cross sections as well as in longitudinal sections of muscle, specially prepared and examined at high resolution, delicate cross linkages between thick and thin filaments have been reported where overlap is present.

By various methods it has now been shown that the thick filaments represent aggregates of the muscle protein myosin, while the thin filaments are composed of actin. Myosin can split ATP releasing the energy stored within its high energy phosphate linkages. This energy is partly converted to mechanical work by the overlapping filament mechanism of the myofibril. It is thought that a mechanical force may be exerted in some way at the cross linkages between the thick and thin filaments causing the thin filaments to slide with relation to the thick, leading to shortening of the sarcomeres. During contraction (Fig. 14) the thick and thin filaments remain at constant length but the sliding of the thin filaments between the thick leads to an increase in their area of overlap and a corresponding decrease in the width of the I bands. The A bands, determined by the length of the thick filaments, remain of constant width. Since the thin filaments slide closer to the M line, the H zone narrows and may even disappear in contraction. When the sarcomere is fully contracted there may even be a region of double overlap in the centre of the A band, where thin filaments from either side of the sarcomere overlap each other as well as overlapping the thick A band filaments. When the area of double overlap is seen in cross section twelve thin filaments may surround each thick filament.

Surrounding the myofibrils is the sarcoplasm, equivalent to the cytoplasm of other cells and containing the same basic fine structural features. The mitochondria vary in size, number and configuration in different muscles according to their energy needs, being most complex in muscles on which severe functional demands are placed, such as the rapidly-contracting insect flight muscles. In longitudinal sections of muscle segmentation of the mitochondria corresponding to the sarcomere pattern may be seen, reinforcing the repeating pattern already present in the myofibrils. The mitochondria, the main source of ATP, lie close beside the myofibril, the site of breakdown of ATP. In addition to mitochondria, dense glycogen particles may often be seen in the sarcoplasm. These can be distinguished from ribosomes by their larger diameter, about 300 Å. A small Golgi apparatus and a few ribosomes are often seen in the sarcoplasm at the poles of the nuclei.

The smooth endoplasmic reticulum of the striated muscle cell is a prominent component of the sarcoplasm on electron microscopy. It has a complex regular segmental arrangement which corresponds to the repeating sarcomere pattern of the myofibril. This system, commonly called the *sarcoplasmic reticulum*, has two distinct structural elements, one transverse in relation to the muscle cell, the other longitudinal. The *T tubules* or *transverse tubules*, cross the muscle cell parallel to the cross striations and lie at a constant position in each sarcomere, normally either at the Z line or at the A-I junction. The T tubule is closely related to the surface membrane of the muscle cell, becoming continuous with it in some species so that the narrow lumen, by communicating directly with the extracellular space, may contain extracellular concentrations of ions. The T tubules may therefore relate the whole thickness of the muscle cell to the surface membrane, ensuring that each sarcomere, even in centrally placed myofibrils 50 μ from the cell membrane, is kept in touch with the functional state of the cell surface.

The longitudinal component of the sarcoplasmic reticulum, the true smooth endoplasmic reticulum of the muscle cell, forms a network which winds between the myofibrils and invests each of them with a lace-like sleeve of inter-communicating cisternae (Fig. 15). The longitudinal tubules are also arranged in a segmental

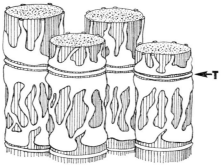

Perspective diagram of adjacent myofibrils in striated muscle

FIG. 15 In this diagram only four of the myofibrils from a single cell are shown. The lace-like pattern of the sarcoplasmic reticulum between the myofibrils is indicated. Note the entirely transverse orientation of the T tubules, T, in contrast to the longitudinal pattern of the other components of the sarcoplasmic reticulum.

pattern, forming expanded foot processes which flank each side of the T tubules. The repeating system consisting of the T tubule with its two closely applied foot processes lying at a constant position in each sarcomere is referred to as the triad of striated muscle. There is no obvious communication at the triad between the T system and the longitudinal elements of the sarcoplasmic reticulum. Thus the segmented sarcomere pattern produced by overlapping thick and thin filaments is accompanied by partial segmentation of mitochondria and by a segmental arrangement of the components of the sarcoplasmic reticulum.

It is known that the contraction of a striated muscle cell is triggered by a wave of membrane depolarisation and repolarisation which is initiated at the motor end plate and spreads along the cell within milliseconds. This is called excitation of the cell. Muscle contraction, the mechanical response to this surface excitation, is a much slower process taking in some cases up to 100 milliseconds to be completed. Contraction involves the chemical and mechanical events which take place within the myofibrils and the sarcoplasm. Although electrical stimulation of the membrane surface of the intact muscle cell causes contraction of the myofibrils, electrical stimulation of the isolated myofibril is not effective. This indicates that the excitation and contraction processes are separate although effectively coupled in the intact cell. It is believed that the sarcoplasmic reticulum is directly involved in this process of *excitation-contraction coupling*.

If a limited portion of the cell surface membrane is depolarised using a hollow glass micro-electrode controlled under phase contrast light microscopy, no contraction occurs unless the electrode is placed at the A-I junction, the point where the T tubule normally approaches the cell surface. When surface depolarisation affects the T tubule, the related sarcomeres contract. This is believed to reflect the segmental function of the T tubule in transmitting surface excitation from the cell membrane to the interior of the cell. In the contraction of the myofibril calcium plays an important but imperfectly understood part. It is thought that the sarcoplasmic reticulum may bind calcium within its cisternae, releasing it on activation of the T system. The presence of calcium allows the splitting of ATP and the triggering of the sliding mechanism. The smooth reticulum might then once again take up the calcium allowing re-establishment of the resting state. This suggested sequence, based on incomplete evidence, presents one possible explanation of muscle function related in detail to the fine structure of striated muscle.

VISCERAL MUSCLE (*Plates 26, 27, 28 and 29 are relevant to this section.*) The biological nature of visceral or smooth muscle is so different from that of skeletal muscle that it is not surprising to find significant differences in structure. Skeletal muscle, designed for rapid, powerful but short-lived mechanical action, has a clearly defined rest length and a narrow range of movement. Visceral muscle on the other hand is slow to respond, contracts with less force, has no clearly defined rest length and may take up different ranges of movement depending on physiological requirements. Moreover, while skeletal muscle depends on its nerve supply for normal activity and undergoes atrophy if this is interrupted, visceral muscle can be isolated from its nerve supply and still retain reasonably normal function. When seen with the light microscope the visceral muscle cell is fusiform, has a central nucleus and shows no striations.

The cell membrane is invested on its outer surface by a delicate cell coat, composed partly of basal lamina material and partly of related collagen fibrils. This barrier lies between each cell and its neighbours except at certain limited areas where their membranes come into close contact. The contact areas at these points have the structure of the close junction, with apparent fusion of the outer laminae of the trilaminar structure of the two membranes concerned. The number of close contacts between smooth muscle cells varies in different sites. They may provide areas of low electrical resistance, allowing the passage of excitation from cell to cell and accounting for the characteristic spread of activity in a sheet of visceral muscle. The term 'nexus' has sometimes been applied to this type of close junction between smooth muscle cells. At other points on the surface of smooth muscle

E

cells numerous membrane invaginations or caveolae form rows of flask-shaped structures. The significance of these vesicles is unknown, but it is thought that they represent micropinocytotic activity.

There is no sarcomere pattern in the smooth muscle cell. The cytoplasm of the cell is filled with quite densely packed filaments of uncertain length measuring up to 90 Å in diameter, generally arranged parallel to the long axis of the cell but sometimes showing whorls and spirals or minor variations in direction. Scattered at random among these filaments are a number of dense patches also of uncertain length, which commonly present cigar-shaped profiles, roughly 700 Å in diameter when seen in cross-section. No meaningful molecular pattern has yet been attributed to the filament arrangement of smooth muscle. At the poles of the nucleus there are cones of cytoplasm free from filaments which contain components such as the small Golgi apparatus, the ribosomes and occasional lysosome-like structures. Mitochondria found both here and at random in the cytoplasm do not show features of high metabolic potential and occupy a small proportion of the volume of the cell. There is no well-organised endoplasmic reticulum. Although there may well be a common metabolic link between smooth and striated muscle at the level of biochemical function, these two types of muscle represent quite different forms of fine structural specialisation for the purpose of contraction.

CARDIAC MUSCLE (*Plates* 26, 27, 31 *and* 32 *are relevant to this section.*) Cardiac muscle, the third main histological and functional type, is present only in the heart. It is required to provide motive force for the circulation by repeated contraction and failure to meet the full demands of the body may be fatal. The muscle of the heart may rest for only a fraction of a second at a time and must be able to give a rapid, powerful, yet sustained contraction. It must have wide resources of power to meet sudden increased demands during contraction, while the action of every part must be co-ordinated so that all effort is directed to the movement of blood and none wasted by inefficient or disorganised contraction.

In its histological appearances, cardiac muscle is distinct from both skeletal and visceral muscle, but combines in its structure certain features of both. The cardiac muscle cell is striated, displaying the same sarcomere pattern as is seen in skeletal muscle. The nuclei, however, are centrally placed. In contrast to the multi-nucleate appearance of skeletal muscle, the nuclei are separated from each other by transverse partitions termed intercalated discs, now known to mark the boundary between individual cells.

The detailed electron microscopic appearance of cardiac muscle resembles closely that of skeletal muscle in a number of important respects. As in skeletal muscle, the cross-striation is due to the presence of a repeating pattern of thick and thin filaments arranged in register and overlapping to form the familiar bands (Fig. 13), the A band, where the thick myosin filaments are found and the I band, formed by the portion of the thin actin filaments not overlapped by the myosin. The Z line bisects the I band, the M line bisects the A band. The action of cardiac muscle is thought to depend upon the same sliding mechanism as that proposed for skeletal muscle, the energy released by ATP hydrolysis causing the actin filaments to be drawn between the myosin filaments with resultant shortening of the sarcomere. The powerful and rapid contraction and the relative constancy of rest length which the striated pattern can provide are well suited to the physiological needs of the heart.

The organisation of the components of the sarcoplasm is not significantly different from that seen in skeletal muscle. The T system and the longitudinal components of the sarcoplasmic reticulum have a similar arrangement in cardiac and skeletal muscle, although in cardiac muscle they may show slightly less regularity. The heart is in constant activity and a high rate of energy production may be called for during prolonged exercise. The mitochondria which provide the source of energy for contraction are correspondingly large and numerous and their cristae are closely packed, pointing to the importance of oxidative phosphorylation in the economy of the cell. Glycogen and lipid droplets, both potential fuel for oxidative processes in the heart, may be found in the sarcoplasm close to mitochondria. The few examples of skeletal muscle which show comparable development of mitochondrial mass have unusual mechanical functions calling also for sustained high energy output.

The *intercalated disc* is the most specific feature of cardiac muscle. Electron microscopy has shown that each disc marks the point of contact between two entirely separate cardiac muscle cells, each with its own nucleus. Cardiac muscle is composed therefore of numerous separate cellular units joined together closely at the intercalated discs. The discs have two separate functions. They provide a mechanical link between cells, preventing their separation during contraction and acting as anchorage points for the contracting myofibrils of each cell. In addition the intercalated discs allow a communication which makes possible the passage of surface excitation from cell to cell. In this way the activity of the heart is coordinated with resulting mechanical efficiency.

The adjacent closely apposed cell membranes at the intercalated disc form specialised areas of contact at different points. There are quite extensive zones where the membranes of adjacent cells appear to fuse, forming an area of close junction between the cells analogous to the zonula occludens of the epithelial junctional complex and the close junction between smooth muscle cells. These areas are thought to allow permeability to ions between cells, permitting the spread of surface excitation throughout the muscle mass. At other points on the disc there are specialisations similar to the desmosome. The insertion of the thin filaments of the myofibril into the cytoplasmic sides of the membranes at these points causes the characteristic density of the disc under the microscope. These portions of the intercalated disc seem to serve the mechanical functions, anchoring the contractile apparatus and providing cell adhesion. Since the disc always replaces the Z line of the sarcomere it lies between two functional units of the myofibril. The zig-zag form of the intercalated disc is characteristic, the occludens areas tending to lie parallel to the myofibrils, the adhaerens areas transverse or oblique.

COMMUNICATION

Nervous tissue is specialised for the storage, processing and communication of information. The basic unit is the neurone, or nerve cell, together with its specialised supporting cells. The nervous tissues of the body are divided into two broad groups, the central and the peripheral nervous system. The central nervous system consists of the brain and spinal cord, the headquarters of neural activity. The peripheral nervous system is made up of the nerves conveying messages to and from the central nervous system, with their associated ganglia.

THE NEURONE (*Plates* 33, 36 *and* 38 *are relevant to this section.*) The neurone is a cell with a compact body which contains the nucleus and typical cytoplasmic organelles, and elongated processes, the axon and dendrites, which extend for variable distances to make functional contact with other neurones. The axon of a nerve cell, although only a few microns in diameter, often less, may be more than a metre in length. The nerve cell cytoplasm is relatively unspecialised in terms of fine structure. The nucleus is not distinctive and the cytoplasm contains the organelles found in other cells. The granular endoplasmic reticulum is well represented. Groups of cisternae of the granular reticulum are found in patches in the cell with moderate numbers of free ribosomes scattered between them. These patches form the characteristic basophilic areas of the nerve cell cytoplasm long recognised by light microscopy as the Nissl bodies. The Golgi apparatus, first recognised in the nerve cell, is of moderate complexity and is scattered around the nucleus without obvious polarisation. The usual fine structural pattern of Golgi membranes can be seen with the electron microscope. Lysosome-like structures are seen and mitochondria are present in moderate numbers.

The cytoplasm within the processes contains relatively few formed structures apart from a few mitochondria. An orientated system of microtubules exists similar to those seen in cytoplasm elsewhere, often called *neurotubules* on account of their site. Neurofilaments, microfibrils of about 60 Å in diameter, vary in their incidence and relative proportions. It is likely that these longitudinally orientated structures in the nerve cell processes are the fine structural basis for the light microscopic 'neurofibrils' which can be demonstrated with silver impregnations. The significance of the 'protein-secreting' apparatus, represented by the Nissl bodies and the complex Golgi apparatus, in relation to neuronal function is not yet clear.

The nerve cell has a high resting *membrane potential* which forms the basis of its excitability and conduction. If this potential is reduced at one point on the cell by physiological or experimental means beyond a critical level, the membrane at this point becomes rapidly depolarised, although it quickly recovers its resting potential through its metabolic activity. A wave of spreading depolarisation is triggered off by this initial localised depolarisation and is propagated across the surface of the cell. The rate at which this wave of surface activity, the *nerve impulse*, passes along the axon depends mainly on the size and character of the axon. In axons of small diameter the rate of conduction of the nerve impulse may be from one to two metres per second, in axons of large diameter it may be over 100 metres per second. In this way information, expressed as the activity of a single cellular unit in the nervous system, is conveyed over long distances.

THE SYNAPSE (*Plates* 32, 33, 36, 37 *and* 38 *are relevant to this section.*) The point of functional contact between nerve cells is called the synapse. The function of the synapse is to effect the transfer of neuronal activity from cell to cell, making possible the 'data processing' functions of the nervous system. Each neurone has synaptic contacts with a great many other neurones, forming a complex network with countless possible functional circuits.

The synapse is a type of contact specialisation peculiar to nervous tissue. Since the nerve impulse passes normally in only one direction along a multi-neuronal pathway, the contacting parts of adjacent neurones which comprise the synapse can be identified as pre-synaptic and post-synaptic components. A narrow intercellular space intervenes between the components of the synapse. There is no cytoplasmic continuity between cells at this point. The pre-synaptic element usually forms a small expansion in which is often found a cluster of mitochondria, suggesting localised metabolic activity. Close to the pre-synaptic membrane lies a group of small membrane-bound *synaptic vesicles* each several hundred Ångstroms in diameter. The intercellular space of 150 to 200 Å between the pre-synaptic and the post-synaptic membranes is known as the *synaptic cleft*. There is often an accumulation of dense material on the cytoplasmic side of both the post-synaptic membrane, and to a certain extent the pre-synaptic membrane. Details of synaptic morphology in different sites are varied, perhaps reflecting differences in function beyond the reach of present neurophysiological investigation.

It is believed that the synaptic vesicles contain transmitter substance which is released into the synaptic cleft when a nerve impulse reaches the pre-synaptic terminal. The transmitter produces localised depolarisation of the post-synaptic membrane which may trigger off a nerve impulse. After its release, the transmitter is rapidly destroyed by enzyme action. In this way the nerve impulse is passed from cell to cell in a chain of neurones, through the release of a chemical transmitter at the synaptic contacts between the cells. The neuro-muscular junction, or motor end plate, appears to have similar characteristics.

The ability of the synapse to transmit the nerve impulse in only one direction is due to the differences between the pre-synaptic component, which releases the transmitter stored in synaptic vesicles, and the post-synaptic component, which responds to this transmitter. The slight delay in transmission of the nerve impulse across the synapse, a characteristic feature, is accounted for by the time taken for the discharge of the synaptic vesicles and the diffusion of transmitter across the intercellular cleft. This intercellular portion of the nerve pathway is more accessible than the intracellular mechanism of the nerve impulse to drugs designed to interfere with nerve function. Drugs which block synaptic transmission are now important in medicine.

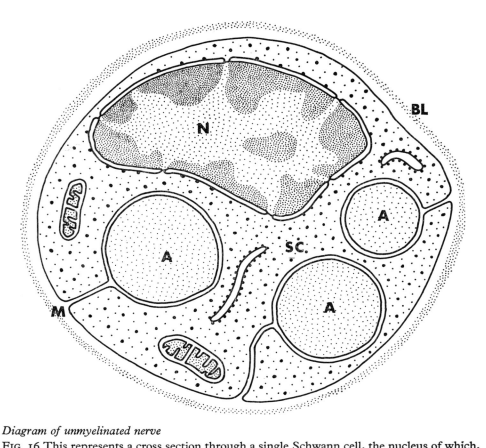

Diagram of unmyelinated nerve

FIG. 16 This represents a cross section through a single Schwann cell, the nucleus of which, N, is included in the plane of section. Three axons, A, of unmyelinated nerves are also cut in cross section. Each axon is surrounded by invaginated Schwann cell surface membrane and is suspended by a mesaxon, M, within the Schwann cell cytoplasm, SC. The axon and the Schwann cell thus retain their individuality while lying in very close relationship. A continuous diffuse basal lamina, BL, lies directly external to the Schwann cell.

NERVE AXONS (*Plates 34 and 35 are relevant to this section.*) A mixed peripheral nerve contains the axons of cells carrying impulses with motor, sensory and autonomic functions. The cell bodies lie in the spinal cord, the dorsal root ganglia and the sympathetic chain. The axon, commonly if perhaps imprecisely referred to as the nerve fibre, is associated with a cellular sheath composed of a chain of satellite cells lying end to end, known as Schwann cells. Two groups of axons are recognised by light microscopy, myelinated and unmyelinated. The myelinated axons are distinguished by the presence of a fatty sheath absent from the unmyelinated axons.

The unmyelinated axon has a simple relationship to the Schwann cell which is best seen in electron micrographs showing a transverse section of a nerve. The

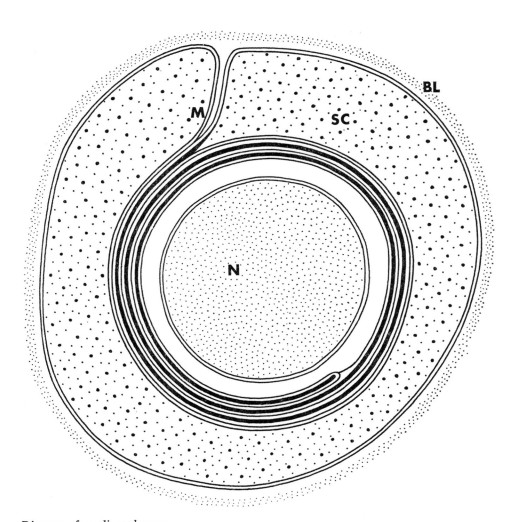

Diagram of myelinated nerve

FIG. 17 A single nerve axon, N, is enclosed within a single Schwann cell, SC. The tri-laminar nature of the cell membrane in each case is shown diagrammatically. The mesaxon, M, appears to become elongated and wrapped around the nerve axon, the myelin sheath being formed by fusion of the membrane elements originating from the Schwann cell surface. The nerve axon thus retains its individuality within the myelin sheath, which itself is a Schwann cell derivative. A diffuse basal lamina, BL, lies external to the Schwann cell.

axon lies in a tunnel within the Schwann cell. It appears to reach this position by invaginating the surface of the Schwann cell, carrying with it a mesaxon of Schwann cell surface membrane. Although the axon is surrounded by the Schwann cell, the surface membrane of each cell is intact. In a typical unmyelinated nerve, a number of axons are carried in this way within each Schwann cell tube, each suspended by its own mesaxon composed of two layers of Schwann cell membrane (Fig. 16).

Since each Schwann cell is much shorter than the axon it surrounds, each axon is enclosed by a succession of different satellite cells along its length. The axon remains shielded by these cells to its finest branches, being naked only when approaching the nerve terminal.

The myelin sheath is recognised in electron microscopy by its marked affinity for osmium. Myelin consists of closely packed regular lipoprotein lamellae with a major periodic spacing of 120 Å and an intermediate minor spacing when seen in thin sections prepared for conventional electron microscopy. The myelin sheath surrounds a single axon, usually of large diameter, forming a tightly packed spiral pattern. During the development of a myelinated nerve, a single axon, having initially the appearance of an unmyelinated nerve, is enclosed within a single Schwann cell sheath. The myelin appears to be formed by rotation of the Schwann cell with relation to the axon. As a result the mesaxon which suspends the axon within the Schwann cell cytoplasm, becomes wrapped round and round the axon. The resulting close-packed double layers of Schwann cell surface membrane fuse tightly together to produce the periodic myelin pattern (Fig. 17). As the myelin sheath gains thickness, the cytoplasmic components of the Schwann cell are pushed peripherally to form a thin rim on the outside of the myelin where the Schwann cell nucleus can also be found. Myelination is apparently an active cellular process, involving a dynamic relationship between the sheath cells and the axon. The fully formed myelin sheath is still part of the living satellite cells, structurally distinct from the axon which it encloses. The process of demyelination initially affects the myelin sheath, and therefore the Schwann cells, rather than the axon.

Since each Schwann cell encloses only part of the length of the axon, the myelin sheath is the product of many Schwann cells, lying end to end. The nodes of Ranvier are the junctions between adjacent segments of myelin, marking the extent of successive Schwann cells. At the node the axon is virtually unshielded and comes into contact with the tissue fluid, while at other points the axon is insulated by the myelin lamellae. In terms of function, the myelin sheath is an important specialisation. Myelinated axons, normally of large diameter, conduct the nerve impulse more rapidly than unmyelinated ones, since activation of the nerve jumps from node to node where the axon is exposed.

THE CENTRAL NERVOUS SYSTEM (*Plates 36, 37 and 38 are relevant to this section.*) Apart from revealing the morphology of the nerve cells and their synapses and of the supporting cells or *neuroglia* in the brain, the electron microscope has emphasised the complexity of brain tissue. Within the brain there is no true connective tissue supporting the cells and apparently no significant intercellular space. The neuroglial cells take the place of connective tissue and surround the other elements of the nervous tissue. Nerve and neuroglial cells are so closely packed in normal circumstances that intercellular spaces greater than 200 Å are rarely seen by conventional methods. The packing of cells throughout the central nervous system is as close as that of a compact epithelium. The 150 to 200 Å intercellular 'space' may in fact be filled with components of the external cell coat.

The neuroglial cells may control the passage of materials between the circulation and the nervous tissues. The cerebral capillaries are surrounded by a cuff of neurogial, mainly astrocyte, foot processes which prevent the nerve cells from coming into extensive contact with the blood vessels. Other processes of the neuroglial cells ramify through the adjacent tissue contacting and often surrounding nerve cell processes. The neuroglial cells may take the place of a connective tissue 'space', providing a diffusion channel for metabolites which is under direct control by the cell. Through metabolic selectivity the neuroglial cells may regulate the nutrition of the nervous system and may constitute part of the physiological blood-brain barrier. Neuroglial cells may even be able to participate in the electrical activity of the brain. Certain neuroglial cells, the oligodendroglia, participate in myelination in the central nervous system and are therefore analogous to the Schwann cells of peripheral nerve, while others, the microglial cells, appear to act as phagocytes after injury. In view of their large numbers and their intimate relationship to the neurones, the neuroglial cells may well have more than a purely mechanical role in the brain.

PHOTORECEPTION
(*Plates* 41b *and* 42 *are relevant to this section.*)
Light is so familiar a component of the environment that its importance can sometimes be overlooked, yet the whole of life as it has evolved on the earth depends in different ways upon the energy of light. In the chloroplasts of green plants the reactions of photosynthesis take place, forming the essential first link in the chain of life on which the more complex animals depend. The animals, for their part, use light to provide information to assist their movement and have evolved photoreceptors, the eyes, which inform the nervous system about the environment. Although the chloroplast and the eye use the energy of light for dissimilar purposes, the components concerned with the trapping and transformation of light energy show ultrastructural specialisations of a surprisingly similar nature.

The chloroplasts of plants are discrete membrane-limited structures lying within the cytoplasm of plant cells, which contain the green pigment chlorophyll. The internal structure of the chloroplast varies greatly in different species of green plant, but their essential common feature is the presence of parallel membrane lamellae associated structurally with the chlorophyll molecules. In the higher plants the lamellae of the chloroplast form a number of discrete specialised packages termed grana, in which the chlorophyll is localised. Photosynthesis can take place only in the presence of chlorophyll associated with the membrane lamellae of the chloroplast, suggesting that the enzymes concerned may be spatially organised on the complex lamellar template. The reactions of photosynthesis use the radiant energy of light to promote the combination of the simple molecules of water and carbon dioxide, resulting in the synthesis of carbohydrate and the release of oxygen. Without this reaction, life as we know it would cease to exist.

The photoreceptor cells of the eyes of animals which are widely separated in the evolutionary scale show essential similarities of fine structural specialisation. In all of them, the light-trapping cells contain numerous membrane lamellae or tubules showing a high degree of spatial organisation, the membranes being associated closely with molecules of photosensitive pigment. The similarity in principle to the chloroplast is clear. In the vertebrate eye the photoreceptor cells are the rods and cones, each of which has an inner and an outer segment joined by a narrow bridge. In the outer segment closely packed parallel membrane lamellae stacked like a pile of coins fill all the available space. Integrated with these lamellae are the molecules of the pigment retinene, closely related to vitamin A, in association with the protein component opsin. It is here that the radiant energy of light is trapped and transduced or converted into cellular activity, leading to the initiation and propagation of a sensory nerve impulse. The inner segment of the photoreceptor cell contains the other cytoplasmic components of the cell including a significant proportion of mitochondria. The bridge between the inner and outer segments is structurally similar to a cilium and is no doubt involved in the functional communication between the two parts of the photoreceptor. The efficiency of the light-trapping mechanism embodied in the retinal photoreceptor is so great that in some cases stimuli amounting to the simultaneous reception of only a few photons are sufficient to generate a visual sensation.

CHAPTER 5

Techniques and Applications

THE ELECTRON MICROSCOPE

The purpose of a microscope is to produce a magnified image of a specimen in order to obtain more information about its structure. Any microscope has three essential parts, a source of illumination, a lens system for producing magnification and some means for observing the final magnified image. The light microscope uses a beam of visible light, generally produced by an electric lamp, to examine the specimen. When high magnification is required, the light from the lamp is concentrated by a glass condenser lens which, on account of its high index of refraction, focusses the light on a small area of the specimen by bending its rays. The details contained in the illuminated specimen are then magnified by a second lens system. The image formed by this system is finally observed through a further lens, the eyepiece. The rays of light which form the magnified image fall directly on the light-sensitive retinal receptors and are registered as a visual image. The specimens most commonly used in biology are ten micron thick sections of fixed tissues or cells, mounted on a glass slide, stained with dyes and examined with transmitted light.

Since a microscope is designed to allow the observation of details not visible to the unaided eye, the most important measure of the efficiency of its performance is the fineness of detail which it permits us to distinguish. This aspect of the performance of a microscope is called its resolving power. The resolution obtained in a microscopic image is determined in figures as the minimum distance between two points which can just be distinguished by the use of the microscope. From theoretical considerations based on the laws of optics it can be shown that the resolving power of any microscope depends essentially on two variables, the wavelength of the light which is used to illuminate the specimen and the numerical aperture of the objective lens, the main magnifying lens in the optical system of the microscope. The glass lenses used in the light microscope can now be produced to exacting standards and lens design does not effectively limit its resolving power. Only the first of these factors, the wavelength of visible light, remains to limit the performance. Visible light, the only form of electro-magnetic radiation to which our eyes respond directly, has a clearly defined range of wavelength which extends from 4,500 Å at the blue end of the spectrum to 7,000 Å at the red end. The theoretical limit of resolving power of a microscope is half the wavelength of the light used to illuminate the specimen so that when visible light is used no details finer than around $0 \cdot 2\,\mu$ (2,000 Å) can be clearly distinguished, however perfect the design of the microscope may be. The limitation lies purely in the physical nature of the light.

It is important to realise that this limitation to resolution which has been described is not simply a limit to the magnifying power of the light microscope. A lens system can be designed to produce any required degree of magnification, simply by making a much larger final image. However, no matter how large the final image is made, no more meaningful information can ever be obtained from it than the resolving power of the system will allow. When the limit of resolution is reached the finest details of the image remain indistinct and further optical magnification will do no more than magnify the blur. Resolving power rather than magnification is the most important measure of the performance of a microscope.

Visible light rays are not the only form of electro-magnetic radiation which can be used to form an image of a specimen, although they are the only form to which the human retina can respond. For example, a beam of electrons can be used instead of a beam of light. In an electron microscope, the wavelength of an electron beam can be as small as 0·025 Å, offering new possibilities in the study of structure. The second variable factor which determines resolution, the design of the lens system, is however so much greater a technical problem than in light microscopy, that even today the best electron microscopes have still not attained the ultimate maximum resolution. Microscopes at present in commercial production offer a resolution of about 5 Å.

The use of electrons presents a number of fundamental practical problems which have now been solved in various ways. Electrons are small negatively charged particles which are very readily absorbed and scattered by any form of matter. It is possible to produce and sustain a beam of electrons only in a high vacuum, since even air alone will scatter electrons. The electron microscope must therefore be a closed system in which a vacuum of at least 10^{-4} mm. Hg must be maintained. The beam is produced within the vacuum chamber by passing a high voltage electric current through a tungsten filament and shielding with metal apertures all but a small pencil of the electrons produced. The electron beam so formed passes through a hollow column surrounded by lenses. Glass lenses are of no value, since glass would absorb electrons. An electron lens makes use instead of the negative charge of the electrons, which allows the beam to be focussed by deflection of the electrons by electromagnetic fields. The lenses of an electron microscope consist of coils of wire wound on hollow metal cylinders designed in such a way that a magnetic field is produced in the centre of the lens by an electric current passing through the coil. The electron beam passes through this field under vacuum as it travels down the microscope column and the extent to which the electrons are deflected can be varied by adjusting the current flowing through the lens coil by means of a variable resistance, controlled by a rotary switch. This adjustment allows the lens to be focussed. By the use of a series of electron lenses of this type, the beam of electrons can be controlled in much the same way as a beam of light in the light microscope. The condenser lens first focusses the electrons to provide a spot of concentrated 'illumination' on a small area of the specimen. A magnified image of the specimen is then formed by a further system of lenses, the objective, intermediate and projector lenses. The image is focussed and magnification altered by varying the currents passing through the different lenses.

The design of the electron lens is perhaps the most difficult technical problem in the production of a high resolution electron microscope and this is the main factor which at present limits the resolution which can be achieved. However, although the microscope can now give resolution better than 3 Å, the specimens used in biological electron microscopy are themselves so imperfect that they limit effective resolution in the study of cells to a greater or lesser extent. Further improvements in the resolving power of the electron microscope cannot be easily used to study cell structure without further improvements in thin sectioning techniques.

It is necessary to convert the electron image formed by the lens system in the electron microscope into a visible form before its information can be registered, since the eye of the microscopist cannot respond directly to the electron beam. By projecting the electron image on to a fluorescent screen at the foot of the microscope column the energy of the electrons is transformed into visible light due to excitation of the chemical coating of the screen. Thus a dense area in the original specimen which scatters electrons will appear as a dark area in the magnified image on the viewing screen, while an area with little dense material in the sections allows the electrons to pass through, so that they reach the screen and produce a bright patch of light. Areas of the section which scatter electrons are described as 'electron dense'. In this way the patterns of density present in the specimen are translated into patterns of light and darkness which constitute the image on the screen. The screen can then be viewed directly by the observer with the unaided eye, or with a binocular light microscope, which magnifies the details on the screen by a further ten times.

A permanent record of the image can finally be made on a photographic plate or film, which responds to electrons as well as to light. For this purpose, a camera is placed under the screen of the microscope. When the screen is tilted from its position in the beam the electron image falls instead on a photographic plate. After the plate has been exposed, it can be removed from the microscope and developed and fixed in the usual way. It then becomes a negative which can be printed in the form of a black and white photograph, the electron micrograph. In practical terms, the micrographs are the main permanent record of material studied by the electron microscope. The convenient light microscopic slide, quickly re-examined and easily marked and shown by projection, makes photographic recording of light microscopic material slightly less important.

Since the maintenance of a high vacuum in all parts of the electron microscope is essential for its operation, a vacuum pumping system is an integral part of all machines. A simple rotary pump is used to clear most of the air from the column of the microscope, while high vacuum pumping is completed by diffusion pumps. To maintain a high vacuum, it is essential that the microscope should be free from air leaking into the column from the atmosphere. Every section of the microscope and every moving control must be sealed to prevent leaks, usually by the use of rubber rings between adjoining metal surfaces. Another technical feature of microscope construction is a water cooling system for the pumps and the lenses, since in both of these places it is important to prevent overheating. In some cases a circulating pump, refrigeration plant and filter system forming a closed circuit are included in the design of the microscope. In modern machines, miniaturisation and transistorisation of electronic systems have helped to reduce the size of complex control circuits, while automation is helping to make routine maintenance and operation more simple in many respects.

ROUTINE SPECIMEN PREPARATION

The use of electrons in microscopy imposes particular demands not only on microscope design, but also on specimen preparation. Since electrons are scattered and absorbed so readily, the ten micron thick sections of tissues used for the light microscope are too thick for electron microscopy. Sections about 500 Å in thickness are now routinely obtained, due to the use of new embedding media, which are less volatile and give more support to the tissue than wax, and new microtomes which are more delicate in their construction and operation. Equally important are the new methods of fixation which have been developed. While the standard histological fixatives preserve the structure of cells well enough for routine light microscopy, the preservation is not satisfactory when more critical examination of the material is possible at higher resolution. The details of cell structure which are made visible by the electron microscope can be preserved successfully only by careful control of the pH, osmotic pressure and ionic concentration of the fixative.

FIXATION It has been found that the most useful general fixative for electron microscopy is a solution of osmium tetroxide, often known as osmic acid, used alone or following prefixation with another fixative such as glutaraldehyde. The deposition of osmium in the tissues is an important factor in the enhancement of contrast, which is a problem in biological electron microscopy because of the thinness of the sections. The high atomic weight of osmium ensures that any structures in which osmium is deposited during fixation appear dense in the electron image. Such structures are described as osmiophilic. The fixative must contain dissolved salts to bring the osmotic pressure and ionic composition to the level found in normal tissues, in order to avoid osmotic damage during fixation. Sucrose, a common addition to fixing fluids, allows the osmotic pressure to be increased without affecting ionic composition. The structure of the cell is damaged by uncontrolled acidity during fixation, which can be prevented by buffering the fixative to a pH of 7·4 with a suitable buffer system. The commonest fixatives use either phosphate, veronal acetate, S-collidine or cacodylate buffer systems. Good preservation of cell structures is also achieved by preliminary fixation with a buffered glutaraldehyde solution followed by postfixation in 2 per cent osmium tetroxide. Suitably buffered solutions of formaldehyde can be used to advantage for special purposes, such as for fixing tissues prior to certain types of electron histochemical procedures.

To be effective, fixation must begin as soon as possible after the circulation of blood to the tissue has ceased. When specimens are being obtained from a laboratory animal, the tissue can sometimes be fixed by dropping the fluid directly on to the organ which has been exposed at operation during anaesthesia. Alternatively the tissue can be removed and chopped immediately into small pieces in a drop of fixative to ensure rapid and complete penetration of the fixative into the centre of each small piece. It is commonly accepted that blocks of tissue for electron microscopy should always be smaller than 1 mm. cube, especially when using osmium-containing fixative, which penetrates tissues slowly and with difficulty. The use of perfusion fixation is more satisfactory, particularly for small experimental animals in which accurate dissection of tissues can be difficult and time-consuming. In perfusion fixation, the fixative which is passed under pressure through the blood vessels of the anaesthetised animal replaces the circulating blood and comes into immediate and intimate contact with the tissues. The material can be dissected at leisure once it has been fixed in this way. In all forms of fixation it is now customary to use chilled solutions to avoid as far as possible unwanted chemical extraction of components of the cells. It is also important that tissues should be handled at all times with the maximum possible care, avoiding crushing and using only the sharpest of razor blades during removal.

EMBEDDING After fixation, normally complete within one to three hours, the pieces of tissue are washed in buffer solution or water and then prepared for embedding. Wax embedding methods used for histology are not suitable for electron microscopy since wax does not support tissue sufficiently for sectioning and is also volatile, evaporating in the high vacuum of the microscope when exposed to the heating effect of the electron beam. Plastic materials have now been developed which do not have the same defects. Since these materials are not usually miscible with water, prior dehydration of the tissue is essential. This is achieved by passing the material through increasing concentrations of ethanol, finishing with absolute ethanol. A clearing agent such as xylol, miscible both with alcohol and with embedding medium, is then used to provide a transition to the final stage of the processing.

The plastic usually takes the form of a liquid monomer solution, which is sufficiently fluid to penetrate through the tissue within a period of several hours. The tissue is placed in a small gelatin capsule filled with the liquid form of the embedding medium. The process of polymerisation is then induced by heating in an oven to temperatures up to 60°C. The polymerised form of the embedding medium is hard and durable, and the tissue is surrounded and infiltrated by this supporting material, forming a convenient block which is ready for sectioning. The process of polymerisation alone may take 48 hours or longer so that processing tissues for electron microscopy takes up to three or four days. Examples of embedding media are the now little used acrylic resins such as methacrylate, familiar as perspex, the more popular epoxy resins, Araldite and Epon, and polyester materials such as Vestopal. Water-soluble media are occasionally used for special purposes.

SECTION CUTTING Electron microscopy demands sections of about 500 Å from these hard blocks, since the electron beam is unable to penetrate thicker specimens sufficiently to allow details of structure to be clearly seen and there is confusion from overlapping images. Ordinary microtome knives are not suitable for cutting

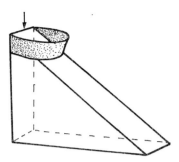

Diagram of glass knife

FIG. 18 The triangular glass knife, a couple of centimetres in height, is shown with tape attached to its shoulder to form a trough behind the cutting edge which is arrowed. The sections when cut float on the surface of the fluid placed in the trough.

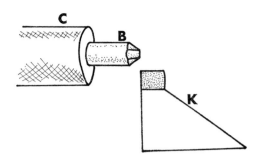

Diagram of cutting position seen from the side

FIG. 19 The knife, K, seen in side view, has its trough attached. The block, B, is clamped in the chuck, C, of the ultramicrotome. During the cutting cycle, the chuck falls slowly, allowing a section to be cut from the trimmed end of the block. The cycle is repeated to form a ribbon of sections.

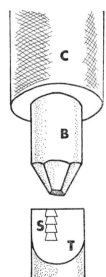

Diagram of cutting position seen from above

FIG. 20 The chuck, C, and block, B, lie above the knife edge. A ribbon of four sections, S, lies on the surface of the fluid in the trough, T. The pyramidal face of the block can be seen.

such thin sections since it is impossible to polish a metal edge finely enough to obtain good sections and glass or diamond knives are used instead. Glass knives are made by breaking triangular shapes from plate glass strips, either manually, using glass cutters and pliers, or mechanically. The sharp edge of the freshly broken angle of a glass knife is extremely fragile and is often badly damaged after only a few sections have been cut. Diamond knives which are also available are more durable and can be used for serial sectioning and for cutting dense tissues such as bone.

Once the knife has been made it is provided with a trough which allows a fluid reservoir to lie behind the cutting edge so that sections, once cut, will float on the surface in a ribbon. The troughs are clipped on to the shoulder of the knife and the edges where the bath meets the knife are sealed by dental wax. The fluid which fills the bath is a mixture of acetone and water.

Before sections are cut from the block, the gelatine is first removed by soaking in water and the tissue is exposed for sectioning by trimming the embedding medium with a razor blade to form a small pyramid on the end of the block, the apex of which is occupied by the tissue. A small flat face is then cut from the tissue, exposing a rectangular area about 1 mm. square, from which the sections will be cut. These and subsequent operations are carried out under binocular microscopy for accuracy of manipulation. The block is fixed in the chuck of the microtome and preliminary thick sections are removed to polish its face. A 2 μ thick section is then mounted on a glass slide for light microscopy. This section may be stained with methylene or toluidine blue or may be examined unstained by phase contrast microscopy in order to localise areas of interest for subsequent electron microscopy. The area of the block face can then be trimmed down to include only those structures of interest in the electron microscopic investigation.

The microtomes used for thin sectioning must be designed to strict tolerances. The machines may be either manually operated or automatic. The knife with its bath in place is clamped in the microtome and the block is fixed in a moving chuck. The block must advance towards the knife at each successive stroke by only 500 Å, so that care is called for in the operation of the microtome. As the sections are cut they float from the knife edge in a continuous ribbon on the surface of the bath. When viewed in reflected light, the colour of the section as it is cut indicates its approximate thickness. Satisfactory sections appear grey or silver. Slightly thicker sections, still useful for some purposes, are gold in colour. Compression introduced during sectioning is removed by exposing the sections to chloroform or xylol vapour.

Tissues for electron microscopic study are mounted for examination by spreading the sections over the surface of a finely perforated copper grid, about 2 mm. in diameter. In this way small areas of the section are stretched between the metal cross bars of the grid which give sufficient support to the section without the need for an intervening layer equivalent to the glass slide in light microscopy. As a result, scattering of electrons by unnecessary layers is eliminated since the electron beam need only pass through the section. The sections are placed on the grid by laying the grid on the ribbon floating on the surface of the bath. The sections adhere to the grid by surface forces.

CONTRAST ENHANCEMENT If sections prepared as described above are examined in the electron microscope without further treatment, the inherent contrast in the tissue is so low that little detail can be distinguished, mainly because the sections are so thin. In light microscopy, contrast can be enhanced by the use of coloured dyes which are bound by chemical reactions to different parts of the section,

F

producing areas distinguished by their different colours. Colour, however, is not a property of the electron image. Contrast in the electron microscope relies instead on the differential absorption and scattering of electrons which depends on the presence in the section of material of high atomic weight. Fixation of the tissue with osmium solutions adds some contrast, but in general contrast must be further increased by exposing the section to a solution of a heavy metal salt, such as lead hydroxide, acetate or citrate, uranyl acetate or phosphotungstic acid. The heavy metal is taken up preferentially by membranes, granules and other distinctive structures, causing them to scatter electrons more than before. Such components appear dense in the final image. It is usually necessary to post-stain in this way to gain sufficient contrast for the adequate investigation of fine structure.

Contrast also depends on the operating conditions of the microscope. If a small aperture is placed in the column at the objective lens the contrast is increased, while the same effect is obtained by operating the microscope at a lower voltage than the usual level of 60 kV. Unfortunately the use of lower voltage increases the heating effect of the beam and damages the specimen more quickly. Operation at high voltage allows the beam to penetrate thicker specimens, but is accompanied by lower contrast. It is possible that high voltage microscopy, which causes little heating effect and little specimen damage, may offer a possible method for the study of living cells by electron microscopy in the future, if problems of low contrast and difficulties of specimen preparation can be overcome.

SPECIAL TECHNIQUES FOR ELECTRON MICROSCOPY

TRACER TECHNIQUES The thin sectioning technique used for transmission electron microscopy is the standard method of specimen preparation for most purposes in biological laboratories. However, in recent years a number of techniques have been developed which have extended the range of the electron microscope beyond the limits of pure morphology. Most of these techniques have evolved from existing light microscopic procedures modified to meet the special demands of electron microscopy. An early example is the use of the silver and osmium impregnation methods in conjunction with electron microscopy for the demonstration of the Golgi apparatus. The position of the electron dense heavy metal deposit allows accurate identification by electron microscopy of the cytoplasmic structures which represent the Golgi image in the light microscope.

By using electron-dense colloidal particles as markers, a number of interesting studies of cell behaviour can be undertaken. The processes of phagocytosis by different cells can be studied by following the progress of fine particles of colloidal thorium or gold injected into the circulation. The individual colloidal particles, too small to be visible directly by light microscopy, are immediately obvious in the electron micrograph, on account of their density. The particles can be seen adhering to the surface of macrophages and their segregation and isolation within cytoplasmic lysosomes can be followed. The extent to which intestinal cells ingest particulate matter from the gut at different stages in their development from new-born to adult in experimental animals has been studied with similar marker methods. The molecule of ferritin is particularly valuable in this context since it is a biological substance of macromolecular dimensions, but has sufficient intrinsic density on account of its high iron content to be directly visible, with characteristic morphology, by electron microscopy (*Plate* 13). The ferritin molecule has been used to demonstrate continuity in frog skeletal muscle between the T tubule and the extracellular space. The living muscle is placed in saline containing ferritin for a period of time and is then processed for electron microscopy. Control muscle cells placed in saline alone have no ferritin molecules within the T system, while the accumulation of ferritin within the T tubules of muscle exposed to the experimental solution indicates that free passage exists in this case between the extracellular fluid and the lumen of the tubule.

A form of direct labelling has been developed for light microscopy to follow immunological reactions in certain circumstances, using antibody labelled with fluorescent dyes. Fluorescent coupling can be accomplished without affecting the biological reactivity of the molecule of antibody. When the labelled antibody is placed in contact with cells or tissues containing its specific antigen the fluorescent marker becomes fixed at the site of the antigen-antibody reaction, where its presence can be detected by ultraviolet light microscopy. In a similar way molecules of antibody can be coupled with ferritin and their position after the antigen-antibody reaction has taken place can be determined by direct electron microscopic examination. The position of the ferritin then marks the location of the antigenic material in the cell. Ferritin labelling can detect the presence of antibody molecules in the endoplasmic reticulum of the plasma cell, and can be used to pinpoint antigenic sites on the surfaces of bacteria and viruses.

AUTORADIOGRAPHY Another technique adapted from light microscopy is the method of autoradiography. The localisation of radioactive material in a section of tissue is made possible by spreading a thin layer of photographic emulsion over the section. The emulsion in contact with an area containing radioactive material becomes exposed by the action of the radiation, causing the formation of a latent image localised to the position of the emitting substance. When the emulsion is then developed and fixed as in conventional photography the areas exposed to radioactivity appear as dense clusters of silver grains overlying the histological section. The position of the grains in relation to the underlying histological details indicates the site of the radioactive tracer and the number of grains in a cluster gives a rough indication of the strength of the source of radioactivity and the amount of radioactive material present in a given area.

By autoradiography, using the DNA precursor thymidine labelled with the radioactive isotope of hydrogen called tritium, it is possible to identify the cells which are engaged in the synthesis of DNA at a particular time, thus providing the basis for research into the kinetics of cell division and cell population renewal. The uptake of radioactive sulphur and its conjugation with mucus has been shown, by autoradiography of the goblet cell, to take place in the Golgi apparatus. Labelled carbohydrate molecules have also been shown to become incorporated in newly synthesised mucus within the Golgi apparatus, whereas radioactive amino acids are incorporated into the protein component of mucus within the endoplasmic reticulum at the base of the goblet cell.

CYTOCHEMISTRY Light microscopic histochemistry and cytochemistry are based on reactions in tissues which can be studied by the use of coloured dyes. In some cases, such as the test for mucin or the PAS reaction, the dye concerned reacts directly with specific chemical groupings in the material which is to be demonstrated. The coloured dye becomes bound to the areas of the specimen containing the material of interest, which can then be localised by light microscopy. In other cases enzymes present in the cell can be demonstrated indirectly by the reactions of dyes with the products of their specific enzymatic action. Enzyme histochemistry has become an important branch of light microscopy, since it allows an insight into cell function at the molecular level. The function of a cell is related to the enzymes it contains and abnormalities may be reflected in diminution or absence of enzymes. There have been many efforts to adapt these cytochemical reactions for use in electron microscopy, in order to localise more precisely the position of the enzymes in the cell in relation to the known components of fine structure.

The nature of electron microscopy imposes restrictions on cytochemical procedures. Suitable markers must be found for electron microscopic localisation. The colourful dyes on which much of light microscopic histochemistry is based are not suitable as stains for electron microscopy. They are organic molecules of low density, visible by their colour on light microscopy, but invisible under electron microscopy on account of their low electron scattering power. A coloured image can not be produced by electrons. The most suitable reactions for electron histochemistry are those which can be adapted to produce precipitates of heavy metal salts at the reaction site. Among the reactions which have been demonstrated successfully by electron microscopy are the phosphatase group, including acid and alkaline phosphatase techniques. The use of the acid phosphatase method has made possible the ultrastructural localisation of the enzyme in sections of tissue, allowing a tentative identification of lysosomes without the need for biochemical studies of cell fractions. Alkaline phosphatase has similarly been localised to the surface of the microvilli in the small intestinal epithelium.

CELL FRACTIONATION The biochemist and the cell biologist are often concerned with the chemical activity of fractions of cell homogenates. Various components of the cell may be isolated from homogenates by the use of the ultracentrifuge and their biochemical behaviour and enzyme content analysed. Along with this it is important to have electron microscopic identification of the structural components present in the fractions. The significance of particular enzyme assays in, for example, a mitochondrial fraction, can be increased by the demonstration of a pure fraction containing little contamination from other sources in the cell. Electron microscopy confirmed the apparent biochemical differences between the main bulk of the mitochondrial fraction of liver cells and a small subfraction which was distinguished by its heaviness and by the presence of hydrolytic enzymes. In this way the biochemical concept of the lysosome was given a structural meaning in agreement with experimental observations. The isolation and electron microscopic identification of pure ribosomal fractions and of fragments of endoplasmic reticulum has made possible the study of different aspects of protein synthesis. The isolation of intestinal cell brush borders and the investigation of enzymes located in this region of the cell have been aided by the confirmation of the purity of the fraction in structural terms.

SHADOWING AND NEGATIVE STAINING Special techniques for increasing specimen contrast have been developed to make possible detailed examination of materials without resorting to thin sectioning. These methods are particularly valuable for the investigation of small particles such as viruses. Virus particles can be examined by placing a drop of a suspension of the virus on a specimen grid covered with a thin carbon or collodion film for support and allowing the suspending fluid to dry. The individual particles, however, are of such low intrinsic contrast that they cannot be effectively studied by direct transmission microscopy without some additional staining procedure.

The size and structure of small particles can be determined to some extent by shadowing the specimen with heavy metal. The specimen grid is placed in a vacuum chamber and metal is evaporated on to its surface at an angle by passing an electric current through a filament. The metal particles are deposited as a fine dust over the surface of the film forming a layer opaque to electrons. However, any structure lying on the surface of the specimen grid during the shadowing procedure gathers a deposit of metal on one side and casts a 'shadow' where it shields the surface from the particles. If the angle of shadowing is known, the size and shape of the object can be calculated from the dimensions of its shadow.

Negative staining provides another method for the investigation of the structure of small particles. The virus can be suspended in a solution of phosphotungstic acid or some other negative staining material and a drop of the suspension placed on a grid and allowed to dry. A dense deposit of stain is left lying on the surface of the film but is present only to a lesser degree over the surface of the virus particles. The stain does however penetrate into very fine surface recesses in the virus, outlining delicate structural details in individual particles. The result of this form of staining is the production of a negative contrast image, the virus particles appearing pale against a dense background. Negative staining is of value in the study of many types of macromolecule, since individual molecules can be seen directly without fixation or embedding artefacts.

THE ELECTRON MICROSCOPE IN BIOLOGY
The first electron microscope was constructed in the early 1930s in Germany and it was not long before production models were available which could resolve details of structure beyond the range of the light microscope. Although the development

of the electron microscope was delayed by the Second World War even then it was some time before this new technique made a significant impact on biological science. There were two main reasons for this delay. The first was the technical complexity of the instrument, which demanded of its operator greater than average skill in engineering and electronics. The operator of an early electron microscope was more often a physicist than a biologist and most of his time was spent in repairs and maintenance of the machine. The second reason was that the techniques for histological specimen preparation for light microscopy were not suitable for the new demands of the electron microscope.

Between 1948 and 1954, however, a number of technical advances in specimen preparation were introduced, including the use of glass knives, plastic embedding, ultramicrotomes and isotonic buffered fixatives containing osmium tetroxide. At the same time microscopes became simpler to operate and to maintain. As a result, there was a sudden breakthrough in the study of biological fine structure. The progress in this field since the middle of the 1950s has been little short of spectacular. In a few years, the ultrastructural foundations of the new science of cell biology have been laid and a revolution has taken place in our concepts of the organisation of living things. The electron microscope has now become useful in many different fields of biological research.

The first result of fine structural study in biology was the resolution by indisputable morphological evidence of many of the controversies which for years had enlivened histology and cytology. The cytoplasm was shown to have a complex and variable fine structure and the true existence of the Golgi apparatus was confirmed. The intestinal striated border was shown to consist of microvilli. A continuous epithelial lining was demonstrated in the pulmonary alveoli. The intercalated disc of cardiac muscle was revealed as a zone of cell contact, disproving the syncytial theory of cardiac muscle. The cell membrane was shown to have a true structural identity and the myelin sheath was found to arise from it. In these and many other cases the results of fine structural study were the final answer to many years of dispute.

Through studies of specialised cells of different types in different species the morphologist has become aware of associations between the newly described subcellular 'anatomy' and the metabolic functions of the cell. The relationship of the granular endoplasmic reticulum in many different situations with the function of protein secretion and the variations of mitochondrial structure in cells with different energy needs are examples of this type of association. Through biochemical studies aided by electron microscopy the significance of these observations has been confirmed and amplified.

Another aim of fine structural study has been the perfection of new techniques and the establishment of standards of tissue preparation in relation to familiar cell types. In this way the merits of epoxy resin embedding and glutaraldehyde fixation were made clear. The establishment of the range of normality in cell structure is not only valuable itself as a means of gaining insight into the nature of the living cell, but is also essential in setting the baseline for experimental and pathological research. The importance of research and development of techniques is still great, since even now the thin sections used in biological electron microscopy are too thick to permit the attainment of top resolution from the electron microscope. The microscopes themselves are now much better than the methods of specimen preparation.

The study of abnormal structure is part of the science of pathology. Since the middle of the nineteenth century when cellular pathology began to develop as a separate discipline, the light microscope has played an increasing part in research and subsequently in the diagnosis of disease. For this reason, the scope of modern diagnostic pathology has been determined by past research using light microscopic techniques.

Since diagnosis can be made only on the basis of recognised abnormalities determined by research, the place of the electron microscope in this important branch of practical pathology is not yet clearly defined. It will however be determined in the future by the insights gained by fine structural research into pathological processes at the cellular level. Already the electron microscope is an important aid to the study of renal glomerular pathology, making possible here as elsewhere a more rational approach to the structure of the tissues affected by disease. In other fields of pathology the current interest in fine structure suggests that a possible future role will be found for the electron microscope in the diagnosis of previously unrecognised forms of cellular pathology.

The value of the electron microscope has already been proved in the study of viruses and viral disease, where the infecting organism is often so small that only electron microscopy can show its presence directly. The classification and structural description of viruses owes much to fine structural studies using negative staining techniques. In practical terms, the electron microscope offers more rapid diagnosis in cases of suspected smallpox and a reliable means of distinguishing this disease from less serious conditions which resemble it clinically.

The horizons of biology, medicine and pathology have been widened greatly in the past by the use of the light microscope. In many respects the electron microscope, less than 40 years after its invention, is making a similar impact. Today however the electron microscope and the light microscope have complementary roles in biological research and are in no way in competition with each other. Just as the light microscope extends the range of the unaided eye, so the electron microscope extends the range of structural study beyond light microscopic limits. Each new technique gathers new information and has its application to particular aspects of any problem. While light microscopy is essential for the study of tissue architecture, the electron microscope is needed to study the fine structure of the cell. In due course there comes a point where the techniques of physical chemistry have more to offer in the study of structure than those of electron microscopy. There are many methods of investigating biological structure and each must be used with discretion in suitable problems if its full potential is to be realised.

CHAPTER 6

Critical Examination of the Electron Micrograph

INTERPRETATION

The two essentials in the interpretation of an electron micrograph are an adequate basis of theoretical knowledge and a systematic technique for 'reading' the micrograph. Full understanding of the image, to the limits of the available theoretical knowledge, is possible only after examination of all of the available details of fine structure, but the information contained in a micrograph is not always fully utilised by the unskilled observer. As a guide for the beginner to 'reading' the electron micrograph, the following points should be considered.

MAGNIFICATION It is essential to establish the approximate magnification of an electron micrograph since this determines the general level of structural detail which it will be possible to see. The magnification of a micrograph may be quoted in two ways, both of which are used in this book. Firstly, the numerical value of the magnification may be given in the caption. Secondly, the micrograph may be marked with a scale which represents a given length at the particular magnification. These two methods of giving information are inter-convertible, since the length of the scale line is determined directly by magnification.

The commonest scale used in biological studies is the micron (μ). The scale line shown in Plate 1 represents one micron, so that parts of the cell can be measured directly by comparing them with this scale. If this scale is measured, it will be found to be about 32 mm. in length. This means that any structure one micron across in the original specimen will measure 32 mm. in the micrograph. This implies a magnification factor of 32 mm./1 μ. Since 1 mm. is equivalent to 1,000 μ, this magnification can be expressed as 32 × 1,000/1 or 32,000. For this reason, the magnification of Plate 1 can also be expressed as a number: 32,000 times. If the scale line represents 0·5 or 0·1 μ instead of 1 μ an appropriate factor of 2 or 10 respectively must be included to obtain the correct magnification. On the other hand, if the magnification is stated only in figures, the length of a 1 μ scale can be determined in millimetres by dividing the figure, 32,000 ×, by 1,000, giving 32 mm. This can then be referred to the micrograph and dimensions calculated by simple proportion. The other unit of measurement used in electron microscopy is the Ångstrom unit (Å).

Since 1 μ = 10,000 Å, it follows
that 0·1 μ = 1,000 Å
and that 1 mμ = 10 Å.

In a very high magnification micrograph, it might be convenient to indicate magnification by a scale line of 100 Å. At a magnification of one million times, a figure within the reach of electron microscopy, a scale line of 100 Å or 10 mμ would measure 10 mm.

The caption, in which the magnification is usually quoted, gives other information which should be noted before the micrograph is examined. Details of the species, organ, cell type, experimental data, fixative, embedding medium and stain may all be given and may be relevant to the interpretation of the micrograph.

RELATIVE SCALE Magnification on its own is meaningless unless the observer is able to relate this to a more familiar standard. It may be helpful to classify micrographs into low, medium and high magnification groups, noting in each case how much of the cell or tissue can be seen. With practice it is possible to give an approximate magnification to an unknown micrograph on the basis of the general appearance of the specimen. In certain cases, 'standard' cell components can be used as built-in scale lines. The thickness of a membrane, the size of a ribosome, the periodicity of a collagen fibre will help to establish the relative scale of the other components which can be seen.

In a low magnification electron micrograph, parts of several cells can be seen and their inter-relationships may give a guide to the nature of the tissue. Several nuclei may be seen in one plate and the cytoplasmic substructure may not be completely clear, although mitochondria, endoplasmic reticulum and secretion granules will be seen. Membranes appear at low magnification as single dense lines, their trilaminar structure not being resolved. In many cases, the distinctive architectural features may be clear enough to allow positive identification of the site of origin of the material as in the low power micrographs of lung (*Plate* 19) and kidney (*Plate* 39).

At medium power only parts of a cell or adjacent cells may be seen, with perhaps parts of an adjacent nucleus. Cytoplasmic components are clearly seen at this level of magnification but the relationships between cells are not always made clear. At medium power, a few mitochondria may be seen, their internal structure being clearly distinguished. Details of lysosome construction (*Plate* 14) or of the sarcomere pattern and sarcoplasmic reticulum of skeletal muscle (*Plates* 26, 27) are easily seen in medium power micrographs.

A high magnification electron micrograph shows only a small part of a single cell or of the contact surface but gives a clear view at high resolution of the detailed structure of a single cell component. Cell type and tissue of origin can rarely be determined from such a micrograph, although a guide may be given in some cases by a distinctive specialisation. Plate 2 shows the trilaminar structure of a single membrane and the formation of a close junction and the tissue is, on this basis, probably epithelial in origin. The close contact between cells in Plate 3 also indicates epithelial identity. The granular endoplasmic reticulum seen in Plate 7 could come from many types of protein-secreting cells. A common technical feature in high magnification plates is a granularity in the background, seen for example in Plates 2 and 18.

PRELIMINARY IDENTIFICATION At low magnification, the shapes and outlines of the cells, the extent of the intercellular space and the presence of intercellular materials are clues to tissue identity. The distinctive anatomical features of the main types of tissue, epithelium, nerve, muscle and connective tissue, make it possible to reach a tentative identification from a preliminary examination of general structure, before details of cytoplasmic organisation in the individual cells are considered.

In epithelial tissues the cells are usually regular in shape and orderly in arrangement. Cell contact is extensive and only a narrow intercellular space is present in many cases. At the base of the epithelium a distinct basal lamina is present, below which collagen fibres may be seen. Intercellular material, however, is not found between the individual epithelial cells. Surface epithelial cells or epithelial cells lining glands have a structural polarity, with a basal surface related to the basal lamina and a free apical surface. Most epithelial cells have surface microvilli which are most elaborate in cells with absorptive functions, but which are found in a more rudimentary form in almost all cells of this type.

The commonest form of connective tissue, loose connective or areolar tissue, has a distinctive open appearance on examination with the electron microscope. The cells are scarce by contrast with an epithelium and are widely separated. Contact specialisations are rare even when anatomical features of the tissue force closer associations between connective tissue cells than usual. Between the cells lie bundles of collagen fibres, distinguished by their characteristic morphology and periodic structure. The presence of collagen, by definition a component of connective tissue, may help to decide the nature of an area in which few other positive identifying features are found. The irregular outlines of the connective tissue cells may often lead to confusing 'plane of section' effects, with small islands of connective tissue cell cytoplasm related only to surrounding collagen. The nucleus of the cell concerned often lies in another plane. Other pointers to connective tissue identity are the presence of small blood vessels and unmyelinated nerve axons enclosed in their Schwann cell sheath. These components are easily recognised in an electron micrograph, their basal laminae forming their boundary with the connective tissue ground substance. The specialised forms of connective tissue, bone and cartilage, are distinguished by the high density of their matrix which surrounds both cells and fibres, as well as by their special cytological features.

In the case of muscle and nerve, recognition is usually made on the basis of distinctive special features such as the sarcomere pattern, the layer structure of myelin, or the close packing of numerous processes in the central nervous system. In muscle and peripheral nerve, the basal lamina possessed by each cell or unit is in contrast to the shared basal lamina of an epithelium. In this way the general features of a micrograph may give a clue to the identity of the tissue or organ.

Subsequently it is necessary to examine the cells in detail for specific structural features. Surface specialisations such as microvilli, contact specialisations, pinocytotic vesicles and basal infoldings of the cell membrane may be relevant to the activity of the cell. Elaborate cytoplasmic membrane systems, prominent mitochondria and secretory inclusions should be noted. In this way the structural features of the cells may be related to the known associations between fine structure and metabolic activity which have been outlined in the earlier chapters.

INSPECTION OF THE ENTIRE MICROGRAPH The whole area of a micrograph must be examined, despite a natural tendency to concentrate attention on the centre. Since the information recorded in the micrograph is only a small proportion of that contained in the entire specimen, none of it should be overlooked. It is important also to remember that the full significance of a collection of micrographs may be diminished by unintentional selection by the microscopist. Striking features which make pleasing micrographs may be over-represented in the results of a study, while areas less striking but perhaps more representative may be overlooked. In electron microscopy, where the attainable magnification is so great and the coverage of a tissue correspondingly small, the dangers of error due to selection are considerable.

RECOGNITION OF TANGENTIAL SECTIONING EFFECTS An obscure or unfamiliar image in a micrograph may be the result of an oblique or tangential sectioning effect, produced when the section 'grazes' by chance the surface of a structure such as a nucleus or mitochondrion, or cuts a membrane obliquely. When this happens, a structure familiar in profile may present an unfamiliar 'surface' view and such tangential effects can make recognition of cell components more difficult. In Plate 13b the left hand centriole is cut in such a way that its tubular subunits are barely recognisable. A false impression of discontinuity may be given when a limiting membrane is cut in this way. In Plates 8, 11 and 31, the mitochondria are cut in such a plane that no clear membrane structure is seen at some points. This effect must not be confused with rupture or injury to the mitochondrion. The cell membrane will show its trilaminar construction only if it is cut directly across, and will appear blurred if obliquely sectioned. In Plate 2, areas of trilaminar resolution alternate with areas cut tangentially, where the boundary between the cytoplasm and the exterior is blurred. This appearance does not indicate cell rupture. Such effects are a reminder that an electron micrograph prepared from a thin section represents a single plane of a three dimensional structure and must always be examined with this mind.

In some respects, the unconventional plane of section may give additional structural information which is not otherwise available. The most striking example of this principle is the representation of the nuclear pore in thin sections. In Plates 15 a,b,c, the nuclear pore is shown in conventional 'normal' section, and also in the oblique or tangential plane which permits a 'surface' view of the annulus surrounding the pore. A similar effect is seen in the nucleus in Plates 1, 12a and 13b. When the cisternae of the endoplasmic reticulum are cut tangentially as in Plates 7 and 12a, patterns of ribosomes in their surface can be made out which are not otherwise seen in thin sections.

THE RECOGNITION OF ARTEFACTS

A histological artefact is any appearance or image in the specimen not present in that form in the living state but introduced by tissue processing or some other factor. All light microscopic observations made on fixed and stained tissues are observations of artefact and much of conventional histology is concerned with the interpretation of systematic artefacts such as protein precipitation and the binding of dyes. Despite this, the histological appearance of tissues prepared by conventional means is now taken to represent a tolerably close approach to their true structure. The artefact of histology has come to be accepted because it is reproducible and consistent and because it has proved both meaningful and useful in the more general context of biology and medicine. In the same way, the artefact of electron microscopy has become accepted. It is both reproducible and consistent since the essentials of cell structure have now been confirmed in different species and tissues. It is meaningful and useful in the context of modern biology since it has made possible a more rational approach to the problems of cellular biology and cellular pathology, especially when related to the biochemical functions of the cell.

It is at the level of the highest resolution of the electron microscope that there is most significant doubt concerning the interpretation of the electron image. At this level, precipitation and dehydration of protein may distort molecular arrangements and the final image, which relies on the accumulation of heavy metal during fixation and staining, becomes difficult to understand. There is as yet insufficient evidence at a corresponding level of resolution from other disciplines to confirm or refute theories of image interpretation in terms of molecular patterns. It is debatable whether the structure of a living membrane or a myelin sheath is as clearly defined or as precisely ordered as the electron image suggests, or whether the ribosome exists in life in the form so familiar to the microscopist. At this level there exists a form of biological uncertainty principle, since to observe the molecular patterns of the living cell we must first make them visible by means which may alter or destroy the underlying structures. Such uncertainty, however, does not effect the value of the electron microscope in studies at lower resolution, where images are more easily related to familiar cellular patterns.

Apart from the systematic artefact upon which observation of tissue depends, there are other technical artefacts which may be induced during tissue processing, by faulty technique. Artefact of this kind is unwanted, since it may obscure meaningful detail and add to the difficulty of interpretation. Technical artefact is particularly common in electron microscopy since the resolution of the instrument is so great that imperfections which would remain unnoticed on light microscopy become obvious. In the course of fixation, embedding, sectioning and microscopy there are many possible sources of artefact induced by technical faults which may at times mislead the observer. Here is an outline of some of these artefacts.

ARTEFACT IN SPECIMEN PREPARATION Soon after the circulation is halted, complex biochemical changes begin to take place in cells which lead to cell death and eventual dissolution. There are structural alterations which accompany these changes. The process of breakdown, called autolysis, is arrested by fixation, which although killing the cell, preserves its structural integrity by halting biochemical destruction of the cell components. Fixation normally relies on the precipitation of proteins by chemical means.

In order to anticipate the structural damage of autolysis it is best to initiate fixation of tissues as soon as possible after the interruption of the blood circulation. The optimum results for electron microscopy are obtained by perfusion fixation, which replaces the circulating blood by fixative solution without any form of intervening trauma to the tissue. It is thought that even a short period of cell anoxia may induce significant alterations in fine structure.

Delays in fixation are usually associated with a range of changes in fine structure to which it is difficult to set clear limits. Such changes are commonly termed 'fixation damage'. In general the presence of pale or empty areas within a cell accompanied by clumping of cytoplasmic structures is suggestive of this type of artefact. There is an unacceptable coarseness of detail which will be obvious only after experience of examining the particular tissue. Other indications of damage are broken membranes, swollen and disorganised mitochondria, vacuolated cytoplasm and distended cisternae of the endoplasmic reticulum. The normal fine granular pattern of the nucleus may often be lost, being replaced by irregular patchy aggregations of chromatin with abnormal 'empty' areas. Minimal damage of this type is of course very difficult to distinguish from normal fine structural variation. It has been suggested that delays in fixation may produce their characteristic structural effects on the cell by making its components more sensitive to damage during subsequent processing, rather than by causing immediate structural disruption.

Delay in fixation is not always enough to account for 'fixation damage'. Deviations from the accepted normal limits of fine structure may be induced by a fixative which is of unsuitable osmotic pressure or pH. Particularly damaging are hypotonic and acidic fixatives. The embedding process, particularly with the methacrylate method now seldom used, may itself cause damage to fine structure in well-fixed tissue. This 'polymerisation damage' seems to result from the rupture of fine structural components by chemical cross-linking during the polymerisation process. Damage to fine structure may also be caused by the use of blunt knives and by crushing caused by clumsy manipulation while the tissue is being cut into pieces of a suitable size. If the blocks of tissue are too large, the fixative in which they are immersed may not penetrate to the centre of the block sufficiently fast to arrest the processes of autolysis.

It is important to remember that a certain proportion of cells have reached the end of their life span at a given time in any tissue. These dying cells may show changes of fine structure which are not the result of artefact but are a natural event. In the intestinal epithelium, worn-out cells are shed constantly from the tip of the villus and it is common to encounter structural changes which must not be confused with fixation damage or with the effects of disease.

Other processing faults may lead to technical artefact. Inadequate dehydration, often due to moisture contaminating the absolute alcohol, may result in failure of the embedding medium to penetrate the tissue, which is then poorly supported after polymerisation. Errors in the embedding schedule or in the composition of the resin mixture may result in a block which is too hard, too soft or of irregular consistency, making it difficult to produce thin sections of good quality. The sections may break up, making microscopy impossible, or pits and holes may destroy the details of fine structure.

ARTEFACT IN SECTION-CUTTING The sections which are used for electron microscopy are so thin that defects which fall generally into two types are readily introduced. A poor knife with a rough cutting edge causes scoring of the section, which appears on examination as parallel lines running in the direction of cutting. The edge of the glass knife is so delicate that any touch will damage it and knives must be handled with extreme care to avoid this type of artefact. The second common artefact in thin sections is 'chatter'. This is a regular transverse banding or rippling of the section, parallel to the edge of the knife, often so fine that it is visible only on examination with the electron microscope. This artefact may make the sections worthless at low or medium magnification. Its cause may be found in slackness in the mounting of the knife or the block, or in external vibrations affecting the ultramicrotome. The periodic variation of section thickness which leads to this banded appearance may be aggravated by faults in the consistency of the block.

Compression of thin sections is inevitable during cutting. Secretion granules and other structures which are normally round appear oval or flattened as a result of compression and the other parts of the tissue are equally, if less obviously, affected. Compression may therefore cause inconsistencies in measurements of cell components at high resolution and compression artefact is an important limiting factor in accurate calculations of dimensions of membranes and other structures. Compression artefact can be corrected in part by the use of xylol vapour to expand the sections as they lie on the surface of the trough after cutting.

Contamination of the sections by dirt after cutting is a troublesome source of artefact. Dirt from the trough or a finely particulate or microcrystalline precipitation from the staining solution may obscure details of structure and cause an unpleasing effect. Staining contamination is often so widely distributed over the section that no free area remains suitable for recording. Contamination not only obscures details but affects image stability at high magnification, causing drift and leading to contamination of the microscope column with loss of resolution.

ARTEFACT IN MICROSCOPY The exposure of a section to the electron beam causes considerable heating due to the absorption of energy from the beam. If the beam is focussed too strongly on the section by the condenser lens, the section may stretch and tear. Prolonged exposure to the electron beam burns the surface of the section causing sublimation of the embedding medium and blurs the outlines of the tissue. During irradiation by the beam there is a fine contamination deposited in a layer on the surface of the section, leading to progressive clouding of the specimen. Thus on prolonged examination of a thin section, there is a steady loss of resolution. Many of these artefacts can, however, be reduced by the use of a cooling device.

Two other common artefacts in electron microscopy, drift and astigmatism, may lead to loss of definition in the micrograph. Drifting of the image may be caused by movement of the specimen or by dirt in the column which may accumulate static charge and deflect the beam. Uneven heating of the section due to dirt on its surface, or scores and splits in the section, may also cause movement of the image. If there is image drift, the micrograph will appear blurred, since exposures of two seconds are routinely used for recording the image on the photographic plate. Astigmatism, a defect in the electron 'optical' system of the microscope, causes a fine blurring of the image in one direction. Astigmatism becomes a particular problem at high magnification when small defects in the 'optical' system of the microscope are made more obvious. Astigmatism is commonly caused by dirt on an aperture or lens where it can have an effect on the electron beam.

Failure to focus the image correctly before recording it is a common fault in electron microscopy. At the exact point of focus the image formed by the microscope lacks contrast. Contrast can be increased by putting the objective lens slightly out of focus. When this is done, interference fringes form around the components of the image, giving a false appearance of greater sharpness. The final micrograph, however, lacks resolution and clarity.

The recording of the electron image on photographic plate or film is an important part of electron microscopy. The viewing screen of the microscope which is examined directly by the observer gives poor resolution and the photographic plate serves as the main permanent record of the specimen. Various artefacts may result from faulty exposure of the plate or from damage to the emulsion.

It is clear that the final electron micrograph must be examined with all these possible artefacts in mind, since failure to distinguish them may lead to errors of interpretation. The student will rarely encounter problems of serious artefact, since only those pictures which are technically adequate are normally published. On the other hand it is important that those who may wish to interpret a collection of routine micrographs obtained during a research project should be familiar with the possible sources of technical artefact. Confusion and errors of interpretation can be avoided only by a consciousness of the limitations of the instrument and the specimen at different levels of resolution.

CHAPTER 7

Electron Micrographs

In order to become confident in the interpretation of electron micrographs it is essential to gain the widest possible experience of different tissues prepared in different ways. For this reason no one book can ever provide more than a starting point in fine structure. Careful study of the plates in this section of the book should lay down a basis for the recognition of the principal cell components and the main tissue types, but this limited selection of micrographs should be regarded only as an introduction to the subject and the interested reader must enlarge his experience elsewhere. A number of books can be recommended, including *The Cell* by *Fawcett*, *An Introduction to the Fine Structure of Cells and Tissues* by *Porter and Bonneville*, *An Atlas of Ultrastructure* by *Rhodin*, *Electron Microscopy* by *Causey*, *Electron Microscopy* by *Haggis*, and *Electron Microscopic Anatomy* by *Kurtz*. Further technical information on the practical aspects of electron microscopy and on specialised techniques used in fine structural study may be obtained from *Techniques for Electron Microscopy* by *Kay* and from *Histological Techniques for Electron Microscopy* by *Pease*. All of these books, and others of value and interest, are readily obtainable from scientific booksellers. Since the electron microscope is now recognised in effect as a standard piece of laboratory equipment, the results of fine structural research may be found in almost any specialist journal in the biological sciences. Among the many journals which include a substantial proportion of fine structural work are *Laboratory Investigation*, *Journal of Cell Biology*, *Journal of Ultrastructure Research*, *Journal of Anatomy*, and *Zeitschrift fur Zellforschung*.

The electron micrographs which are included in this chapter are accompanied on the facing page by an extended caption in which both the main and the incidental features are described and commented upon. This section of the book has been designed to stand to some extent on its own but cross references have been included to assist integration between the text and the micrographs. The magnification of each plate is indicated both in the form of a scale line, usually representing one micron unless otherwise specified, and in the form of a direct magnification factor.

The illustration opposite shows a modern high resolution electron microscope, the Philips EM 300, an instrument with many advanced design features. The central microscope column stands in the centre of the desk with the viewing chamber at its foot and the high tension supply cable entering the gun chamber at the top. The operating controls are placed on the desk panels at either side of the viewing chamber.

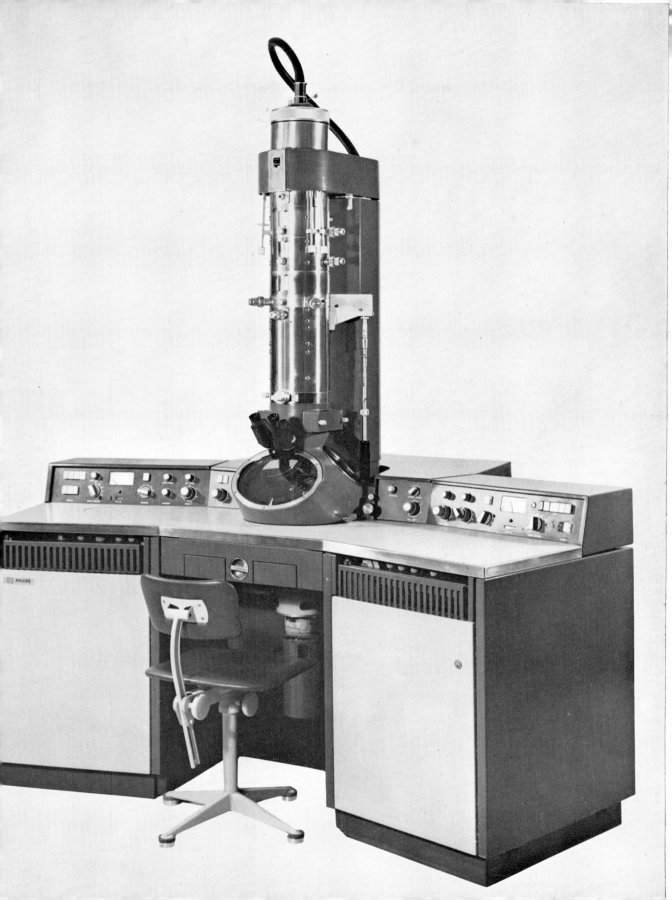

PLATE 1. CELL COMPONENTS

The plasma cell seen in this low power micrograph displays many of the main cytoplasmic components which are commonly encountered in the study of cells. The appearances seen are typical of an active protein secreting cell. The prominent cisternae of the granular endoplasmic reticulum lying in the peripheral cytoplasm are filled with finely granular or flocculent material of appreciable density. This appearance may represent stored antibody newly synthesised in the endoplasmic reticulum. The large Golgi apparatus, formed from several units of Golgi structures, is typical of this cell type. Lamellae, vacuoles and vesicles are all seen, the vesicles being particularly numerous. The mitochondria are not usually prominent.

The plasma cell lying in connective tissue is separated by a significant space from its neighbours. A small portion of a fibroblast is present and a few collagen fibres are seen in the upper left hand corner of the micrograph. The rounded contours of the cell and its solitary appearance are evidence that this cell is an independent unit, rather than part of a gland with many related cells in close contact. The appearance of this micrograph is that of loose connective tissue.

In a number of respects this micrograph illustrates the principles and importance of tangential section effect. The surface membrane of the cell at the point marked C, has been cut obliquely, losing the sharp well defined appearance which is seen elsewhere and which is typical of any membrane. This appearance must not be interpreted as rupture of the cell surface. There is still an area of diffuse appreciable density representing the obliquely sectioned membrane. If the membrane were ruptured the damaged ends would perhaps be seen and the escape of cytoplasmic components from the cell might be observed. The nucleus also shows tangential sectioning effect. It is small, and its outline is diffuse while the nuclear envelope is not resolved clearly in its typical form as seen in other micrographs. This suggests that the main bulk of the nucleus lies outside the plane of section, and the present view is the result of a 'grazing' section through the edge of the nucleus. The oblique angle which this imposes on the membranes of the nuclear envelope makes it impossible to resolve its structure clearly and the perinuclear cisterna is not well demonstrated as a result. The oblique cut however, does permit a semi-surface view of the nuclear envelope to be obtained at its margins. The 'face view' of the nuclear pores, surrounded by their annuli, is obvious at the points arrowed.

The cisternae of the granular endoplasmic reticulum also show areas of tangential section. In the lower right hand corner of the micrograph, the endoplasmic reticulum is present, but its cisternae are cut obliquely. In this way a 'surface' view of the ribosomes is obtained as they lie in association with the membranes limiting the cisternae. Rosettes and spirals of ribosomes are numerous, suggesting that the members of each group may be functionally associated with each other. In this way tangential effects can be used to obtain information not otherwise readily available concerning parts of the cell.

C	Cell membrane cut obliquely
F	Fibroblast
G	Golgi apparatus
GER	Granular endoplasmic reticulum
M	Mitochondria
N	Nucleus
R	Ribosomes
S	Connective tissue space with collagen fibres
Arrows	Indicate nuclear pores seen in face view.

TISSUE Human small intestine. Osmium fixation, lead staining.

MAGNIFICATION 32,000 ×.

REFER TO Plates 7, 9, 10, 12, 15, 16, 21, 33, 47.
Pages 15, 19, 22, 37, 57, 88, 91.

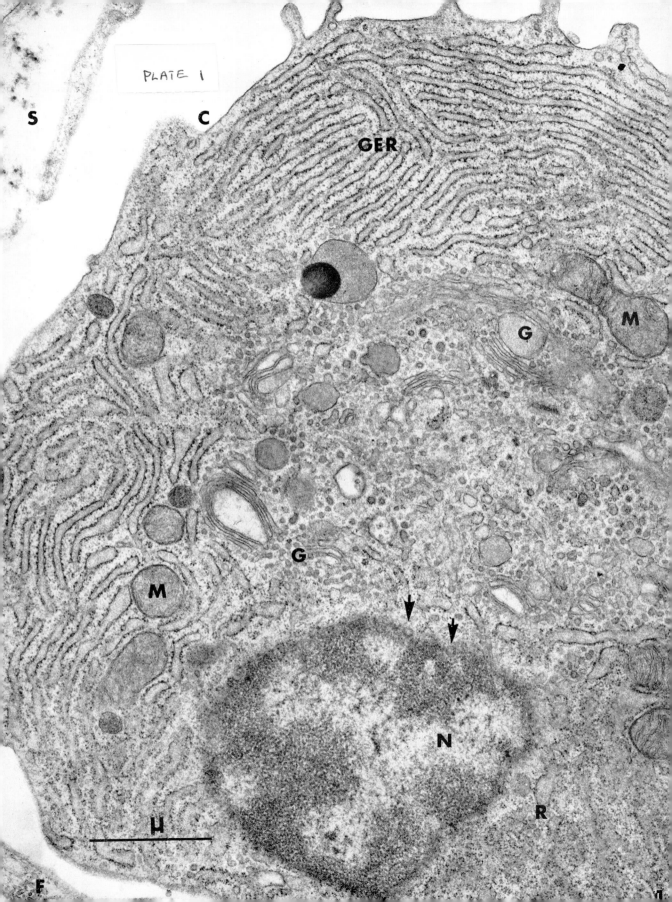

PLATE 1

PLATE 2. MEMBRANES: CELL SURFACE

This is a high magnification micrograph of the surfaces of adjacent seminal vesicle epithelial cells and the gland lumen into which they discharge their secretion. Several stubby microvilli projecting from the cell surface are cut at different angles. Those cut in cross section in the space between the two cells shown appear isolated on account of the plane of section.

The trilaminar pattern of the surface membrane of these cells is clearly seen both at the cell apex and over the microvilli. There are two dense components separated by a narrow interspace of lower density. The 'unit membrane' concept has gained support from images of this kind. At the lower right hand corner of the micrograph the trilaminar surface membranes of the two adjacent cells come together and their outer lamellae fuse to form a close junction or zonula occludens. This specialised area, which extends unbroken like a hoop around the apex of each cell, may act as a seal between the lumen of the gland and the intercellular space, as well as having a possible importance in ionic interchange between cells. The other components of the junctional complex are not included in this plate, since they lie proximal to the zonula occludens, that is off the foot of the page. The zonula occludens, therefore, is formed by coalescence of the membrane structures of two adjacent cells. The trilaminar membrane appearance is seen also in the membranes surrounding two small secretion granules arrowed in the apex of the cell on the right of the field of view.

It is important to remember that the trilaminar appearance of membranes is seen clearly only in high resolution electron micrographs, usually taken at high magnifications. At low magnification and low resolution the membrane at the cell surface appears as a single dense line, and two adjacent separate cell membranes are separated by a gap of 150 Å. The low power 'dense-pale-dense' image of two cell membranes, must be clearly distinguished from the high resolution trilaminar image shown here, which represents the fine structure of a single membrane.

CJ	Close junction
CM	Cell membrane with trilaminar structure
L	Lumen of gland
MV	Microvilli
Arrow	Indicates secretion vacuoles showing trilaminar membranes.

TISSUE Mouse seminal vesicle epithelium. Permanganate fixation, lead staining.

MAGNIFICATION 250,000 ×.

REFER TO Plates 3, 4, 5, 23, 31, 47.
        Pages 4, 10, 37, 89.

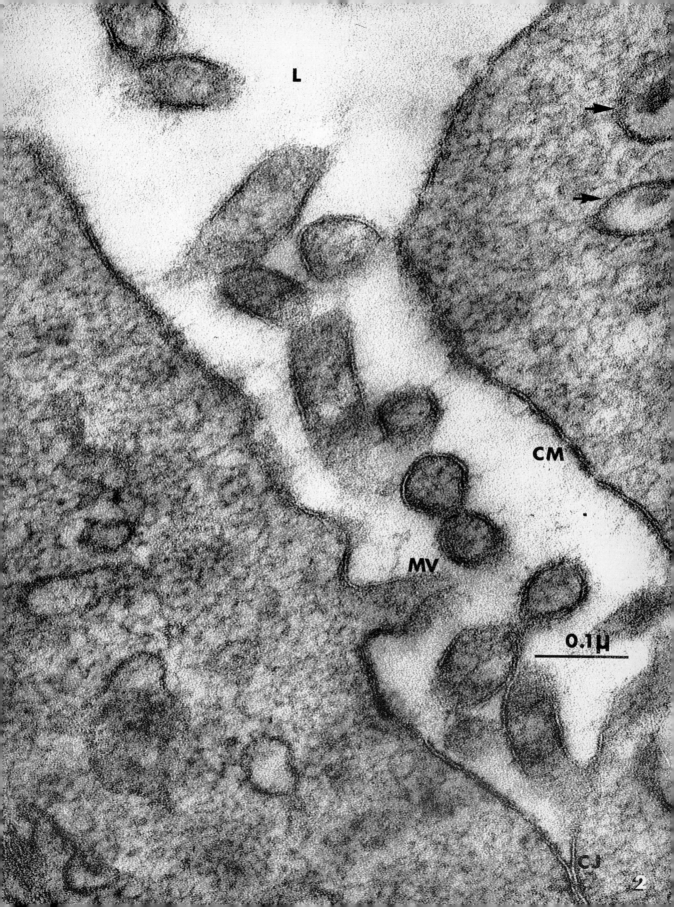

PLATE 3. CELL CONTACTS: DESMOSOME

This is a high magnification micrograph of two adjacent epidermal cells showing a point of adhesion. The complex nature of the large desmosome shown is apparent. The diffuse dense zone on each side of the structure marks the insertion of the cytoplasmic filaments into the cytoplasmic side of the cell membrane of each half of the desmosome. At the upper part of the desmosome the trilaminar structure of each membrane can just be distinguished. A distinct central dense component is seen between the cells as indicated by the arrows. This material may represent part of the cell coat which could have a 'cementing' function concerned with the adhesion properties of the desmosome. The desmosomes and attached filaments in the epidermis suggest intercellular bridges on light microscopy. True intercellular bridges are not, however, seen in the electron micrograph. At the point marked Y the adjacent cells are not held in such close apposition as at the desmosome. Interdigitating processes of cytoplasm cross the narrow pale intercellular space. The desmosome is a discontinuous, button-like structure, as opposed to the continuous 'girdle' formed by the zonula occludens and zonula adhaerens of the junctional complex.

The mitochondria are small and often not clearly seen. Ribosomes are plentiful, however, lying free in the cytoplasm, with little organised granular endoplasmic reticulum. Aggregates of fine filaments are seen throughout the cytoplasm. These have a structural significance shown by their attachment to the desmosomes and they contribute to the mechanical strength and the resistance to attrition shown by the skin. The accumulation of keratin filaments synthesised in these cells is followed by the eventual death of the cell.

Among the other cytoplasmic components present are numbers of smooth surfaced vesicles and a small portion of a cisterna of granular endoplasmic reticulum, in the lower left hand corner of the plate. The large specialised desmosomes linking these cells are an indication of the epithelial nature of the tissue. This feature alone suggests that adhesion between cells is of significance in the tissue, even if its identity were not known. It is not possible however, to attribute any specific function or molecular composition to cytoplasmic filaments on the basis of fine structure alone, biochemical and physiological evidence being required in addition. Cytoplasmic filaments may be seen in cells with many different functions.

D	Desmosome or macula adhaerens
F	Filaments inserted into the desmosome
GER	Granular endoplasmic reticulum
M	Mitochondrion
R	Ribosomes
V	Vesicles
Y	Intercellular space
Arrows	Indicate the midline dense component in the intercellular cleft of the desmosome.

TISSUE Frog skin. Glutaraldehyde fixation, uranium staining.

MAGNIFICATION 82,000×.

REFER TO Plates 2, 4, 6, 30.
Pages 10, 33, 54.

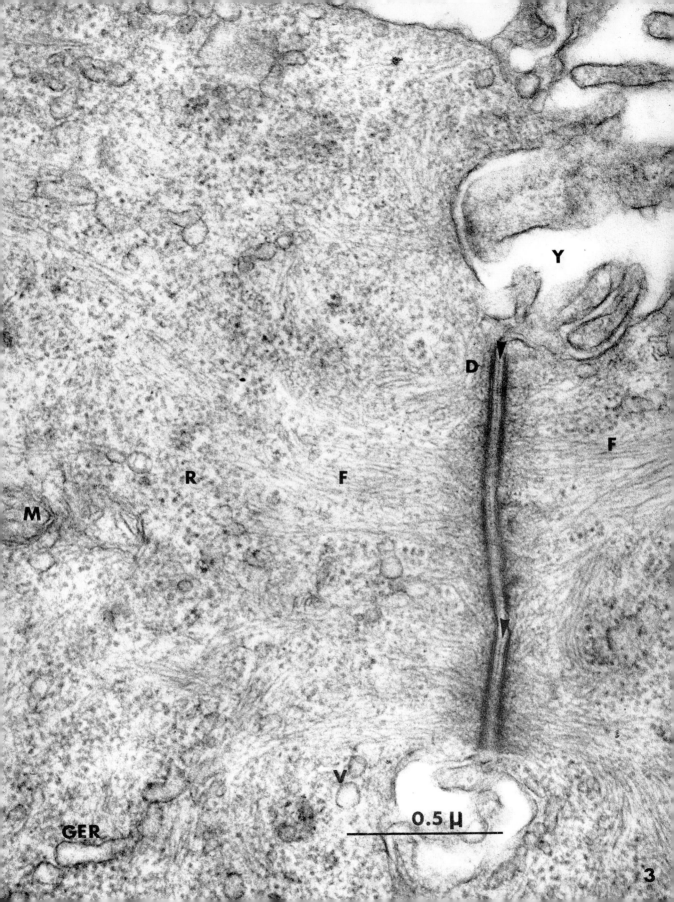

Y

D

F

R

F

M

GER

V

0.5 μ

3

PLATE 4a. CELL CONTACTS: JUNCTIONAL COMPLEX

This is the junctional complex of the intestinal epithelium seen at high magnification, parts of two adjacent absorptive cells being seen. The lumen of the intestine lies to the right of the plate, where the roots of the microvilli, which form the striated border of light microscopy, can just be seen. The apical surface membranes of the two cells fuse at the zonula occludens, or tight junction, forming a seal between the lumen and the intercellular space, the narrow cleft which separates the contact surfaces of the two cells. The zonula adhaerens with its characteristic surrounding area of increased density and its rather wide intercellular gap lies proximal to the zonula occludens, that is, to its left. The desmosome or macula adhaerens is not clearly seen in this case. The remainder of the contact surface shown here is free from distinctive structural specialisation. The pale intercellular space of about 150 Å separates the two cell membranes. The cytoplasmic membrane structures seen close to the cell membrane represent parts of the smooth endoplasmic reticulum.

TISSUE Human intestine. Osmium fixation, lead staining.

MAGNIFICATION 90,000×.

REFER TO Plates 2, 3, 5, 8, 25.
    Pages 9, 10, 44

PLATE 4b. CELL CONTACTS: DESMOSOMES

This micrograph at high magnification shows two desmosomes joining two intestinal epithelial cells. Areas of unspecialised contact surface are also seen. The cytoplasmic filaments inserted into the region and the increased density of the inner surface of the cell membrane at the desmosome can be seen. An intermediate dense line bisecting the intercellular cleft at the desmosome is present, perhaps representing the mucopolysaccharide cell coat with a possible cement function. The unspecialised contact area is marked by arrows, the intercellular gap being about 150 Å wide. No intercellular dense material is present. The appearance seen at the point arrowed where two cell membranes are separated by this constant narrow interspace, must be distinguished from the trilaminar 'unit' construction of a single cell membrane, such as is seen in Plate 2.

Arrows   Indicate unspecialised area of contact surface.

TISSUE Human intestine. Osmium fixation, PTA Staining.

MAGNIFICATION 68,000×.

REFER TO Plates 2, 3, 25.
    Pages 9, 10, 33, 44.

PLATE 4c. CELL CONTACTS: DESMOSOMES

This is an area of contact between two epidermal cells shown at high magnification. Two prominent desmosomes, similar to the desmosome shown in Plate 3 are seen with associated bundles of cytoplasmic filaments. The intercellular dense line can be made out at these points. The unspecialised contact surface, at the point indicated by Y, shows an intercellular space greater than the usual gap of 150 Å. This gap is crossed by processes of the epidermal cells which appear as tongues of cytoplasm bounded by the cell membrane, which at various points such as that indicated by the arrow shows trilaminar construction.

Arrow   Indicates area of trilaminar construction in the cell membrane.

TISSUE Frog skin. Glutaraldehyde fixation, uranium staining.

MAGNIFICATION 62,000×.

REFER TO Plates 3, 6.
    Pages 9, 10, 33, 54

D	Desmosome	Y	Intercellular gap
F	Filaments	ZA	Zonula adhaerens
MV	Microvilli	ZO	Zonula occludens
P	Processes of cytoplasm		

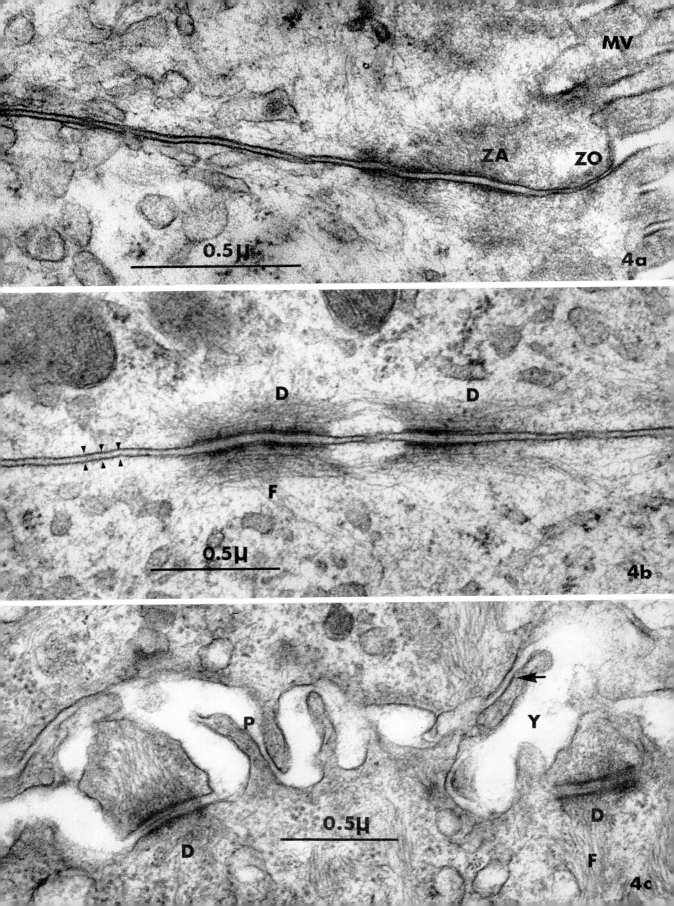

PLATE 5. MEMBRANE SPECIALISATIONS: MICROVILLI

This high magnification micrograph shows the microvilli of an intestinal absorptive cell, cut in cross section. Each microvillus, projecting from the apical surface of the absorptive cell like a finger, consists of a central core clothed by the surface membrane of the cell. When cut in cross section these finger-like processes appear as circular profiles. Since each microvillus is about 0·1 μ in diameter, they are beyond the reach of the resolution of the light microscope as individual structures, appearing visible only when gathered in a closely packed row to form the intestinal striated border. These non-motile processes should be clearly distinguished from the larger, motile cilia shown in Plates 43 and 44.

The trilaminar appearance of the surface membrane covering the microvilli is seen in each of the individual microvilli shown in this cross section. The components of the membrane structure are shown clearly at the point arrowed. The pale interspace between each of the dense laminae of the membrane structure is around 35 Å in width. Although the dimensions vary in different sites, trilaminar structure is present in most biological membranes. In this micrograph, the membrane, which is the digestive-absorptive surface of the intestine, is more specialised than in other sites. This specialisation is reflected in its unusual thickness.

The space, S, between the microvilli is pale, but there is an area of slightly increased density around the outside of each microvillus. At some points it seems as if a radial pattern of diffuse strands of material extends from the outer lamina of the membrane into the space between the microvilli. This appearance is seen for example between '3 o'clock' and '6 o'clock' on the arrowed microvillus, and is thought to represent the fuzzy coat of the intestinal microvilli, not particularly prominent in this species, but at times clearly seen.

The core of the microvillus is filled with longitudinal filaments in which occasionally a suggestion of tubular structure can be detected. In cross section, as in this micrograph, these longitudinal components appear as a closely packed group of dense dots or profiles in the core of the microvillus up to 40 or more in number. They may stiffen the microvillus structurally, or may act as attachments for enzymes or as channels for diffusion. This core is separated by a pale zone several hundred Ångstroms wide from the surface of the microvillus.

The core of the microvillus should be compared with the axial filament complex of the cilium, shown in Plates 43 and 44. Although the microvillus is not a highly organised structure by comparison with the cilium it still shows significant fine structural specialisation which forms the basis for a complex functional and metabolic organisation.

MV	Microvillus
S	Space between microvilli into which surface fuzz extends from the cell membrane
Arrow	Indicates trilaminar membrane structure covering the microvillus.

TISSUE Mouse intestine. Osmium fixation, lead staining.

MAGNIFICATION 250,000 ×.

REFER TO Plates 2, 25, 43, 44.
        Pages 5, 7, 11, 44, 57.

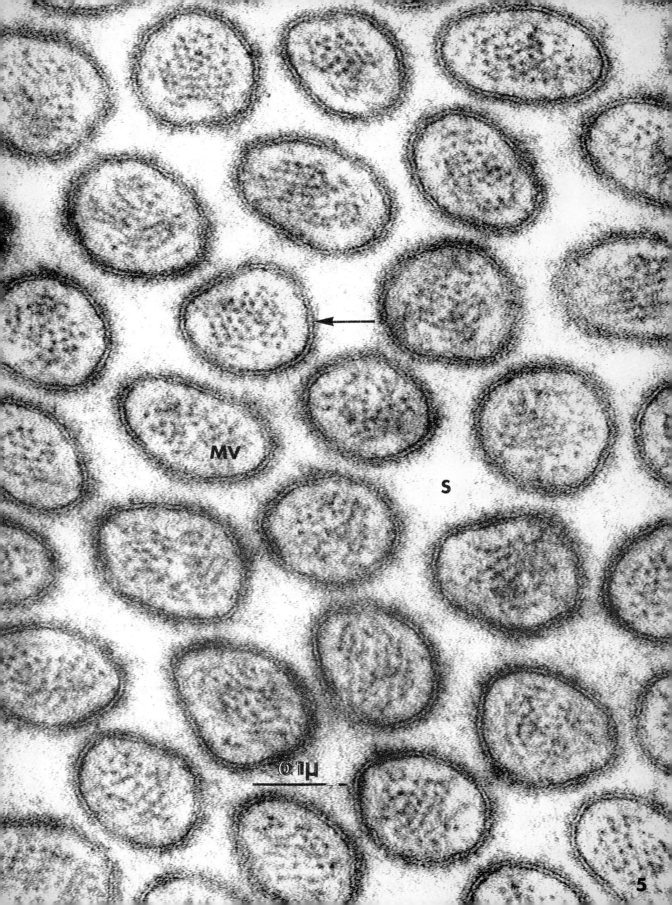

MV

S

0.1μ

5

PLATE 6. BASAL LAMINA

This micrograph shows the interface between the epidermis and the underlying connective tissue at moderate magnification. The epidermal cells lie to the right of the micrograph and the connective tissue to the left. Between them is a discrete basal lamina (or lamina densa), separated slightly from the base of the epithelial cells, but following their irregular contours. The lamina has no distinct organised substructure, but has a diffuse felted appearance which is quite different from the clear-cut image of the cell membrane close to which it lies.

On the connective tissue side of the field of view, the collagen fibres are mostly cut in oblique and transverse sections, giving rise to small dense circular profiles, often closely packed. At times the collagen fibres appear to fuse with the substance of the basal lamina. There are no connective tissue cell processes present in this micrograph.

The epidermal cells pictured on the right of the page are filled with dense fibrillar material. These fibrils manufactured by the skin cells probably consist of keratin, a fibrous protein which has a protective function on the skin. Granular endoplasmic reticulum is not a prominent feature but free ribosomes are plentiful. There are not many mitochondria. At the point marked Y there is an intercellular space crossed by tongues of cytoplasm from each of the cells. This intercellular space, being within the bounds of the epithelial layer, contains no collagen fibres. No special adhesion points are seen in this micrograph, but they are common elsewhere. The curious peg formation seen at the cell base and the interdigitating processes between cells may be related to functional peculiarities of the skin in this species, such as its part in respiratory gas exchange.

BL   Basal lamina or lamina densa
C   Connective tissue space with collagen fibres cut at different angles
F   Fibrils within epidermal cytoplasm
M   Mitochondria
Y   Intercellular space

TISSUE Frog skin. Glutaraldehyde fixation, uranium staining.

MAGNIFICATION 45,000×.

REFER TO Plates 17, 18, 20, 21, 24, 28a, 35, 39, 40.
        Pages 12, 54, 56, 90.

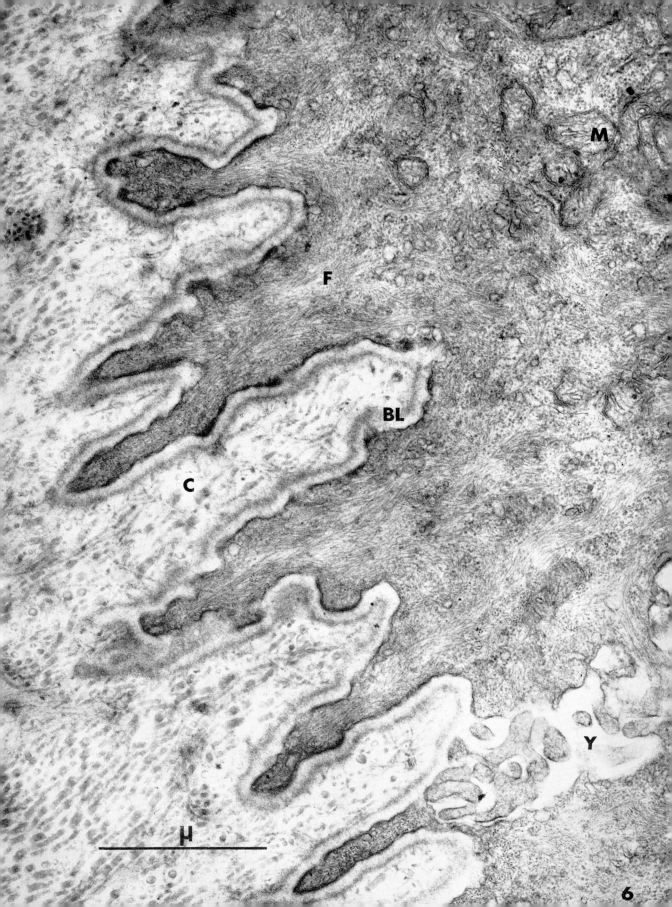

PLATE 7. GRANULAR ENDOPLASMIC RETICULUM

In this high magnification micrograph the elaborate granular endoplasmic reticulum of the protein-secreting plasma cell is shown. The membranes of the reticulum divide the cytoplasm into two phases, the pale cavity of the cisternae, marked X, which intercommunicate extensively, and the denser cytoplasmic ground substance, marked Y, in which the dense ribosomes are found, and in which the mitochondria and other formed cytoplasmic structures are situated. The cisternae appear relatively empty in this plate, since they contain material of low electron density. At times flocculent material suggestive of a protein precipitate can be found in the cavities. At other times in certain types of cell, formed intracisternal granules may appear.

The ribosomes lie not only at random in the cytoplasmic ground substance, but also attached to the outer or cytoplasmic surfaces of the membranes of the endoplasmic reticulum. The inner surfaces are free of specialisations. The characteristic angularity of the ribosomes and their relatively constant size (150 Å) can be assessed from this micrograph. The only other uniform particulate components of cytoplasm commonly encountered in electron micrographs are glycogen, with a particle size of around 300 Å, and ferritin, with a diameter of under 100 Å.

At the upper and lower left hand corners of the field, there are areas where the cisternae have been sectioned obliquely. At these points the normally sharp membranes of the reticulum are blurred as a result of tangential sectioning effect.

M   Mitochondrion
X   Cavity of the cisterna
Y   Cytoplasmic ground substance with ribosomes
Z   Area of oblique section of the membranes

TISSUE Human intestine. Osmium fixation, lead staining.

MAGNIFICATION 78,000 ×.

REFER TO Plates 1, 8, 12, 13, 23, 28a, 33.
       Pages 20, 22, 37, 91.

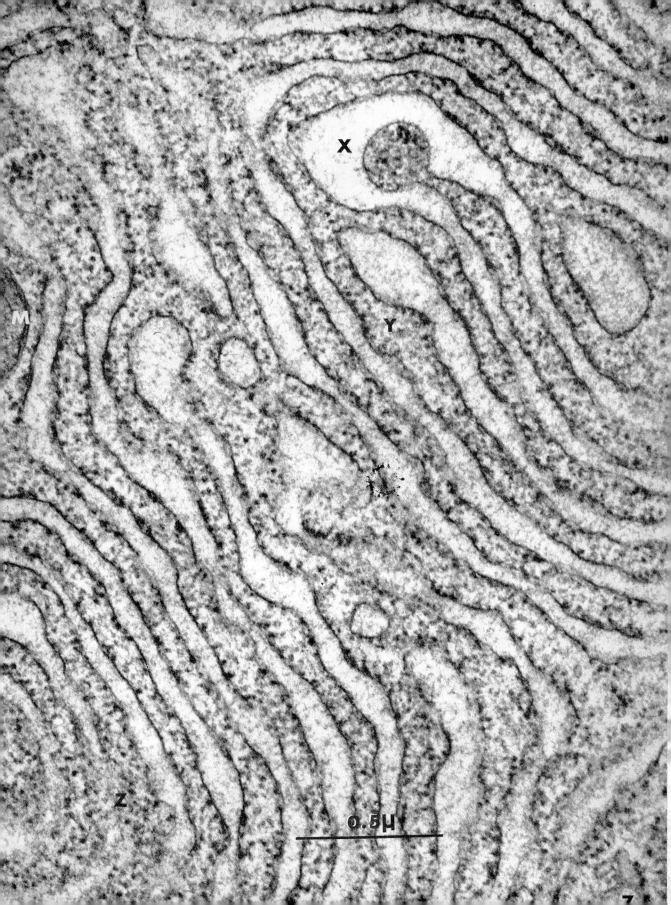

PLATE 8. SMOOTH ENDOPLASMIC RETICULUM

This is a medium power micrograph of the smooth or agranular reticulum from an intestinal absorptive cell. Much of the area of the plate is occupied by closely-packed tortuous cisternae of the smooth reticulum, while on the right hand side of the field other cytoplasmic components can be seen.

Between the dense, sinuous mitochondria, a number of cisternae of granular reticulum are seen. These form a clear contrast with the smooth surfaced profiles, being distinguished by the presence of attached ribosomes. At some points smooth and granular areas appear to be continuous. While granular endoplasmic reticulum is usually associated with a protein-secretory function in the cell, smooth endoplasmic reticulum may have different metabolic functions depending on the type of cell in which it is found. Smooth endoplasmic reticulum is often vacuolar in form and not cisternal or tubular as seen here. Fixation with glutar-aldehyde preserves the tubular form of the smooth reticulum more consistently than fixation with osmium.

The mitochondria are quite plentiful, but relatively small. The matrix is relatively dense and the outer mitochondrial space, which extends into the centres of the cristae, is pale in contrast, giving rise at times to a 'negative contrast' image. Two dense structures in the lower right hand corner of the field have a dense granular content which suggests that they may be siderosomes, cytoplasmic structures containing stored ferritin.

The effect of tangential section of mitochondria can be seen at the point arrowed, where the blurred appearance of the limiting membrane indicates an oblique cut. The clear cut mitochondrial envelope which is normally present is therefore not clearly seen. Such an appearance does not mean that there is rupture of the mitochondrial membrane but is simply a consequence of the plane of section.

F	Ferritin-containing granules or siderosomes
GER	Granular endoplasmic reticulum
M	Mitochondria
SER	Smooth or agranular endoplasmic reticulum
Arrows	Indicate areas where the mitochondrial wall has been obliquely sectioned.

TISSUE Human small intestine. Osmium fixation, lead staining.

MAGNIFICATION 33,000 ×.

REFER TO Plates 7, 11, 13, 21, 25, 26.
        Pages 23, 27, 44, 91.

G

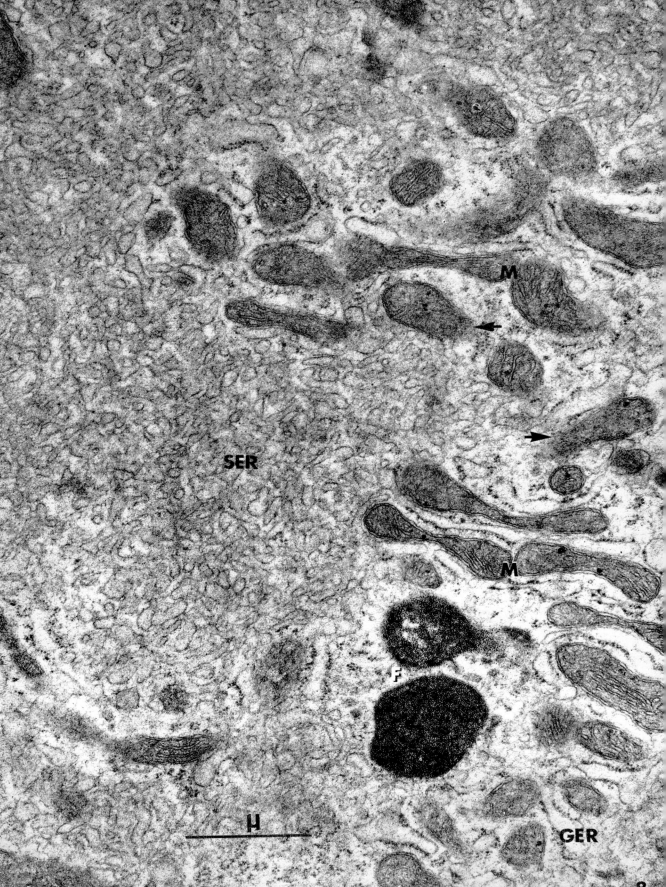

M

SER

F

GER

μ

8

Plate 9a. Golgi Apparatus

This plate shows the Golgi apparatus of a plasma cell at moderate magnification. The three main components of the apparatus are present. The membrane lamellae, comprising a number of closely stacked adjacent sacks, are dilated at points to form the Golgi vacuoles. Surrounding these parts of the apparatus are the numerous Golgi vesicles, membrane-bound structures of small diameter. Several small granules appear to be forming in the centre of the apparatus.

Around the outer aspect of the Golgi apparatus lie the ribosome-studded cisternae of the granular endoplasmic reticulum. When the cisternae are cut obliquely as shown at X, the surface view of the ribosomes attached to their membranes reveals patterns of organisation not otherwise seen. Groups of structurally related ribosomes in spirals and rosettes are commonly seen in such views. There is a mitochondrion present in the picture but it does not clearly display the typical internal organisation. Despite this, its size and the presence of the outer and inner membranes make its identity clear. Since the mitochondrion has been cut transversely the cristae lie parallel to the plane of section and are not clearly seen.

Tissue Human intestine. Osmium fixation, lead staining.

Magnification 54,000×.

Refer to Plates 1, 7, 10, 24, 33.
Pages 20, 22, 25, 38.

Plate 9b. Golgi Apparatus

This plate shows the Golgi apparatus of a developing mucus-secreting cell in the intestine, at moderate magnification. Part of the nucleus appears in the micrograph, and in the upper right hand corner of the field the contact surface between adjacent cells can be seen, the two cell membranes being separated by a narrow interspace. No obvious contact specialisations can be seen.

The Golgi apparatus itself is in the form of a horse-shoe, surrounded by typical cisternae of the granular endoplasmic reticulum. At the inner aspect of the apparatus, secretion granules are seen, indicating a functional polarity of the apparatus, with an 'input' side on its outer aspect, and an 'output' side in its concavity at which granules are despatched. There is occasionally a suggestion that the endoplasmic reticulum is forming buds (arrow) which may be able to detach themselves from the cisternae and pass to the Golgi apparatus, carrying with them packets of material from the cisternae. This would provide a route by which material synthesised in the endoplasmic reticulum could be transported to the Golgi apparatus, where chemical alterations may take place in the secretion and where the secretion granules are finally formed.

Arrow   Indicates an area where it appears a small vesicle is forming by budding from the endoplasmic reticulum.

Tissue Mouse intestine. Osmium fixation, lead staining.

Magnification 64,000×.

Refer to Plates 1, 7, 10, 24, 33.
Pages 22, 25, 38.

CM	Cell membranes in contact
G	Granule
GER	Granular endoplasmic reticulum
L	Golgi lamellae
M	Mitochondrion
N	Nucleus
Va	Golgi vacuole
Ve	Golgi vesicles
X	Oblique cut of endoplasmic reticulum showing rosettes and spirals of ribosomes.

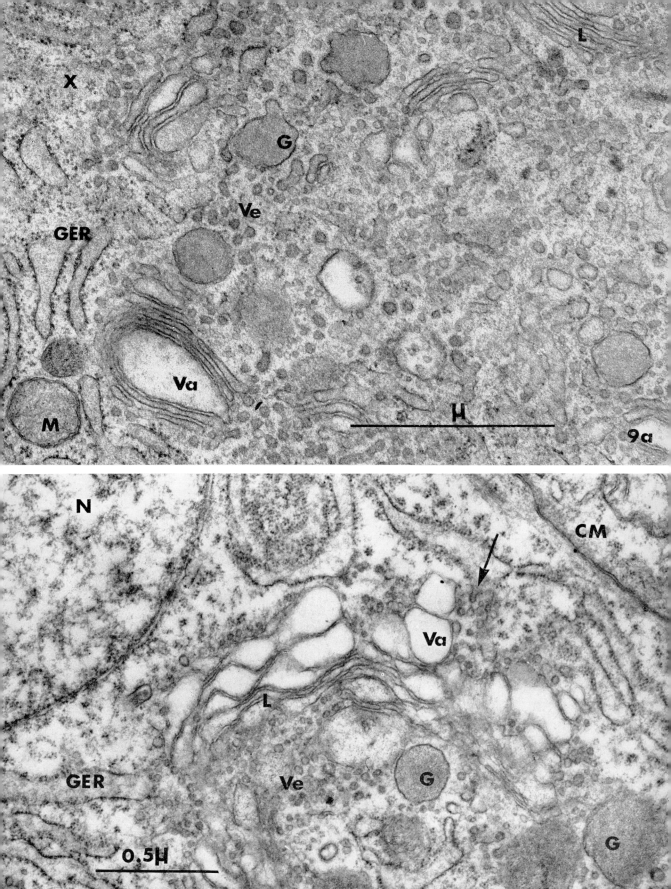

9a

9b

PLATE 10. GOLGI APPARATUS

This is a medium power micrograph of a spermatid showing the Golgi apparatus and its close relationship to the nucleus, which lies on the left of the plate. The nucleus is distinguished by the diffuse granularity and patchy appearance of the nuclear material as well as by the presence of the surrounding nuclear envelope.

The Golgi apparatus has flattened sacs, with only a few dilated vacuoles. Golgi vesicles are seen close to the lamellae and between the cisternae of the endoplasmic reticulum and the Golgi apparatus. At one point, marked by an arrow, there is a suggestion that a vesicle may be forming by budding from a component of the endoplasmic reticulum. On careful inspection the individual membranes of the Golgi apparatus can just be seen to have a trilaminar structure, a feature common to membranes of different types in the cell.

The Golgi apparatus in the spermatid takes part in the formation of a structure known as the acrosome, a cap which is applied to the nucleus and persists in the mature sperm. The acrosome consists of a membrane-limited sac in which material produced in the nearby Golgi apparatus accumulates. This closed sac and its diffuse granular content can be clearly seen in the micrograph. At one point in the acrosome there is a condensation of dense material close to the nucleus which is termed the acrosome granule. The nuclear envelope and the membrane surrounding the acrosome remain separate. It is possible that the acrosome may be a type of lysosome structure, concerned with penetration of the ovum.

The mitochondria seen in this cell are not of the usual configuration as illustrated in Plate 11, but this form is quite commonly encountered in micrographs of the testicular germ cells. Such mitochondria may be particularly sensitive to osmotic changes at fixation or during processing.

A	Acrosome
AG	Acrosome granule
G	Golgi apparatus
GER	Granular endoplasmic reticulum
N	Nucleus
NE	Nuclear envelope
M	Mitochondrion
Arrow	Indicates an area where there appears to be a vesicle forming by budding from the endoplasmic reticulum.

TISSUE Mouse testis. Osmium fixation, lead staining.

MAGNIFICATION 53,000 ×.

REFER TO Plates, 1, 9, 33.
Pages 15, 25, 33, 59.

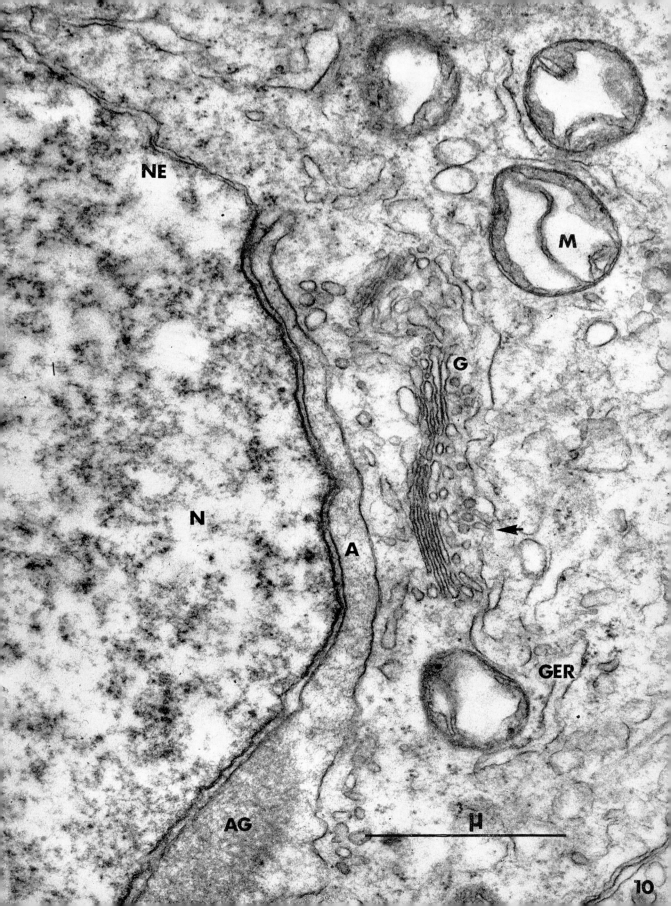

PLATE 11. MITOCHONDRIA

This micrograph shows part of a cell at high magnification. The cell, from the stomach gland of the fowl, has two functions, the secretion of acid and digestive enzymes. The various components of fine structure reflect this double function.

The size and internal organisation of the mitochondria indicate that oxidative metabolism can proceed at a high rate in this cell. The closely packed parallel cristae, extending from side to side of the mitochondrion, are typical of an acid-secreting cell and are seen in other cells with a high rate of oxidative metabolism such as cardiac muscle. In addition, smooth surfaced vacuoles of the smooth endoplasmic reticulum are found, another feature of acid-secreting cells. The vacuolar configuration of the smooth reticulum is perhaps the result of adverse fixation conditions.

Moderate quantities of granular endoplasmic reticulum are also seen in these cells, providing evidence for a zymogenic or enzyme-secreting function. Such components are not prominent in a pure acid-secreting cell such as the mammalion gastric parietal cell. The large dense granule present in the picture is a zymogenic secretion granule. In these cells, in addition to the components shown, there is a well developed Golgi apparatus not seen in the present micrograph. Thus the gastric gland cell of the fowl has the fine structural apparatus for both forms of secretion combined in the same cell. In the mammal these functions have become localised in different cells, the zymogenic chief cell and the acid-secreting gastric parietal cell.

In this section the zymogenic granule has no clear-cut limiting membrane. This is probably due to the plane of section. When zymogenic granules are cut through their 'equator' the close fitting membrane can be distinguished, but when sectioned closer to the 'poles' the surrounding membrane is cut obliquely and cannot be clearly seen on account of tangential effects. Tangential or oblique sectioning effects are also seen at areas marked X. At these points there is blurring of the membranes of the mitochondrion which results from oblique sectioning of the cristae.

GER	Granular endoplasmic reticulum
S	Smooth endoplasmic reticulum in vacuole form
X	Areas of oblique section of mitochondrial cristae
ZG	Zymogen granule

TISSUE Fowl proventriculus. Osmium fixation, uranium staining.

MAGNIFICATION 70,000 ×.

REFER TO Plates 7, 8, 30, 31.
          Pages 23, 27, 37, 40, 42, 91.

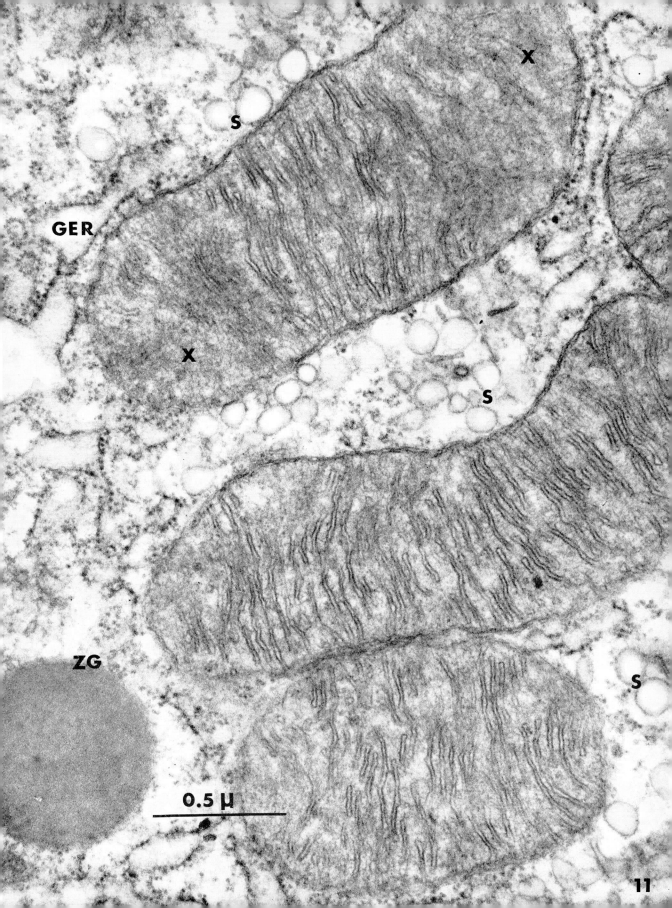

GER

S

X

X

S

ZG

S

S

0.5 μ

11

PLATE 12a. CENTRIOLE

This medium range micrograph shows part of the nucleus of a plasma cell with adjacent cytoplasm. The nucleus shows the dense irregular chomatin pattern commonly seen in plasma cells. On the right of the field a single centriole is seen in longitudinal section, lying close to a typical Golgi apparatus. In this longitudinal section, the cylindrical nature of the centriole can be appreciated but the complex architecture of the components cannot be seen. The association of the centriole with the Golgi apparatus is a common one. It is not unusual to find the second member of the pair of centrioles lying at right angles to the first, but this is not the case in this micrograph.

Several other features of this plate are of interest. The area covered by the nucleus is small in view of the magnification, which suggests that only a small portion of the nucleus is included in the plane of section. This is confirmed by the poorly defined margins of the nucleus. The perinuclear cisterna and the membranes of the nuclear envelope are not clear-cut, indicating that they have been obliquely sectioned. This plate should be compared with Plate 16, in which the nuclear envelope has been sectioned in such a way that the membranes are clearly seen. One result of this oblique cut is the appearance, at points arrowed, of nuclear pores, seen in face view, showing clearly the surrounding dense circular annuli or collars around their outer margins. Tangential effect also accounts for the appearance of the mitochondrion pictured above the nucleus, which has barely been included in the plane of section. The limiting membranes, obliquely sectioned, are not clearly seen. The cristae, however, can still be made out. Some of the cisternae of the granular endoplasmic reticulum are clear-cut, while others, marked X, are poorly defined on account of oblique sectioning of their membranes.

C	Centriole in longitudinal section
G	Golgi apparatus
GER	Granular endoplasmic reticulum
M	Mitochondrion tangentially sectioned
N	Nucleus
X	Obliquely sectioned cisternae of the granular endoplasmic reticulum

Arrows   Indicate areas of oblique cut of the nuclear envelope showing pores with surrounding annuli.

TISSUE Human intestine. Osmium fixation, PTA staining.

MAGNIFICATION 44,000×.

REFER TO Plates 1, 7, 13b, 15, 16.
        Pages 15, 19, 22, 34, 91.

PLATE 12b. CENTRIOLE

In this micrograph a single centriole is seen in cross section. The nine subunits which comprise the centriole are present. Each consists of three tubular components in a characteristic form, while no central elements are present within the centriole. In these respects the centriole differs from the cilium.

TISSUE Mouse intestine. Osmium fixation, uranium staining.

MAGNIFICATION 75,000×.

REFER TO Plates 12a, 13b, 43.
        Pages 15, 19, 22, 34, 91.

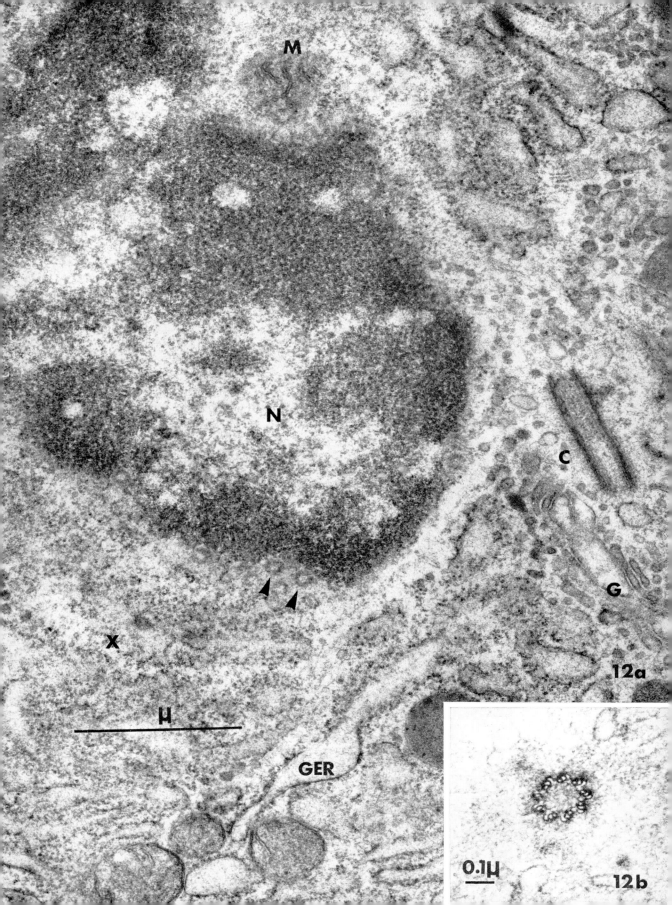

M

N

X

μ

GER

C

G

12a

0.1μ

12b

PLATE 13a. FERRITIN

High magnification view of a portion of cytoplasm from a macrophage showing the presence of ferritin. The numerous small dense particles are individual ferritin molecules, rendered visible by virtue of the high concentration of iron in the core of the molecule. The particles lie both free and enclosed within dense granules which appear to be a form of iron store. A fine substructure which can be distinguished in some of these ferritin molecules may reflect a supposed octahedral pattern of the iron core of the ferritin molecule. The dimensions and morphology of ferritin are sufficiently characteristic to allow these particles to be distinguished from ribosomes and from particulate glycogen. Ferritin is a smaller molecule than either. It can be understood that the ferritin molecules, if attached to antibody as a label, can act as a satisfactory tracer for immunological reactions.

TISSUE Mouse epididymis. Glutaraldehyde fixation, lead staining.

MAGNIFICATION 108,000 ×.

REFER TO Plates 7, 12a, 12b, 15, 16, 28a.
         Pages 31, 54, 83.

PLATE 13b. FERRITIN AND CENTRIOLES

A lower power view of the same cell. The dense ferritin-containing granules, sometimes called siderosomes, are particularly prominent, but ferritin molecules are widely scattered through the cytoplasm. They do not appear within the mitochondria, the Golgi sacs, the endoplasmic reticulum or the nucleus, but several particles are present in the hollow centres of the centrioles, suggesting that these components are open to the cytoplasm. Both centrioles of this cell are present in this section, the one on the left being obliquely sectioned while that on the right is cut in transverse section, revealing the nine subunit construction as seen in Plate 12. Notice the association between the Golgi apparatus, in the upper part of the plate, and the centrioles.

Mitochondria can also be seen in this section, but one of them has been cut obliquely so that its limiting membranes are blurred and the outlines of the mitochondrion are indistinct. Two portions of the nucleus appear in the plate on the left of the field, the upper portion cut tangentially, showing a surface view of nuclear pores, as indicated by arrows, but obscuring the nuclear envelope.

Several 'coated vesicles', thick-walled structures with a fuzzy lining, are seen in this micrograph, one close to the two nuclear profiles on the left of the field, another close to a mitochondrion on the right of the picture. These coated vesicles are thought to be formed by selective micropinocytosis from the surface of the cell.

C   Centrioles
G   Golgi apparatus
Gr   Ferritin-containing granules or siderosomes
M   Mitochondrion
N   Two parts of the nucleus included in plane of section.

Arrows   Indicate nuclear pores sectioned tangentially showing surrounding annuli.

TISSUE Mouse epididymis. Glutaraldehyde fixation, lead staining.

MAGNIFICATION 46,000 ×.

REFER TO Plates 7, 12a, 12b, 15, 16, 28a.
         Pages 8, 31, 34, 54, 83.

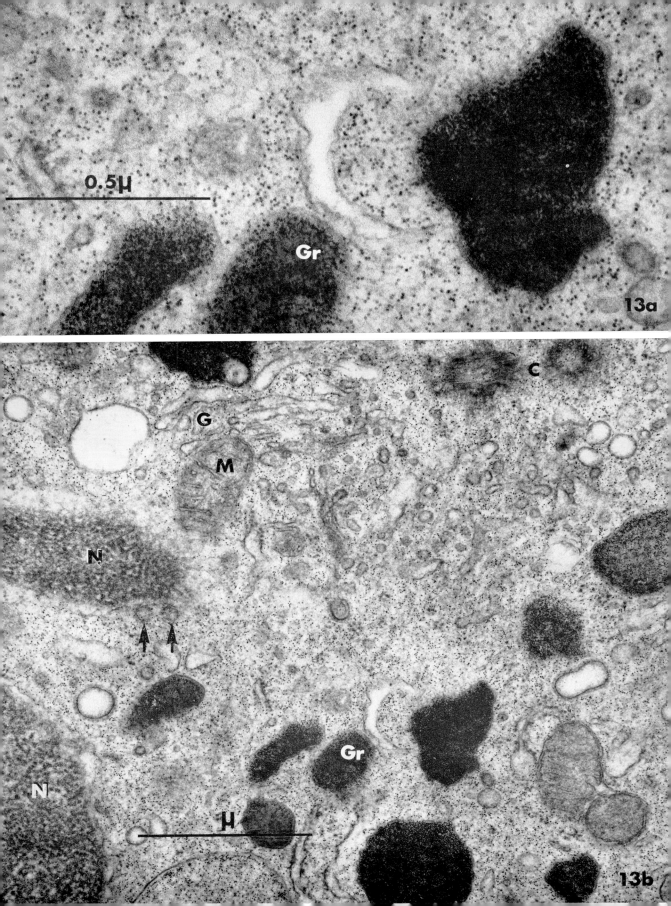

0.5µ

Gr

13a

C

G

M

N

N

↑ ↑

Gr

µ

13b

PLATE 14. LYSOSOME

In this micrograph, at moderate magnification, part of the pulmonary alveolar septum is shown. The cell outlines are difficult to follow. The central large dense structure is a large secondary lysosome lying within the cytoplasm of a pulmonary macrophage. The pleomorphic dense appearance of this structure, with granular areas, homogeneous areas and lamellar components, points to lysosomal identity, although it would be necessary to demonstrate the presence of acid hydrolases within the structure to confirm this view directly. The smaller dense structures, relatively homogeneous in appearance, are probably primary lysosomes rich in hydrolytic enzymes which have not yet participated in the process of intracellular digestion, a specific function of the macrophage. The final structure of the secondary lysosome is dependent on the nature and structure of the material taken up by the macrophage in phagocytosis.

The presence of typical cross-striated collagen fibres close to the surface of the central cell confirms its connective tissue nature. There is, however, little connective tissue space, since the pulmonary alveolar septum is a thin partition.

At the region marked X, there is a contact surface between two cells, without any evidence of specialised contact points. Both these cells are probably macrophages, their close relationship being imposed by their confined physical surroundings.

On the left of the field a basal lamina is arrowed, underlying a dense cell which is probably epithelial in nature. This is likely to be the alveolar lining epithelial layer.

C   Collagen in connective tissue space
L   Lysosome
M   Mitochondria
X   Contact surface of macrophages
Arrow   Indicates a basal lamina or lamina densa underlying a layer of cells with dense cytoplasm.

TISSUE Mouse lung. Glutaraldehyde fixation, uranium staining.

MAGNIFICATION 60,000 ×.

REFER TO Plates 6, 19, 20, 25a, 25b.
          Pages 12, 31, 48, 51, 56, 90.

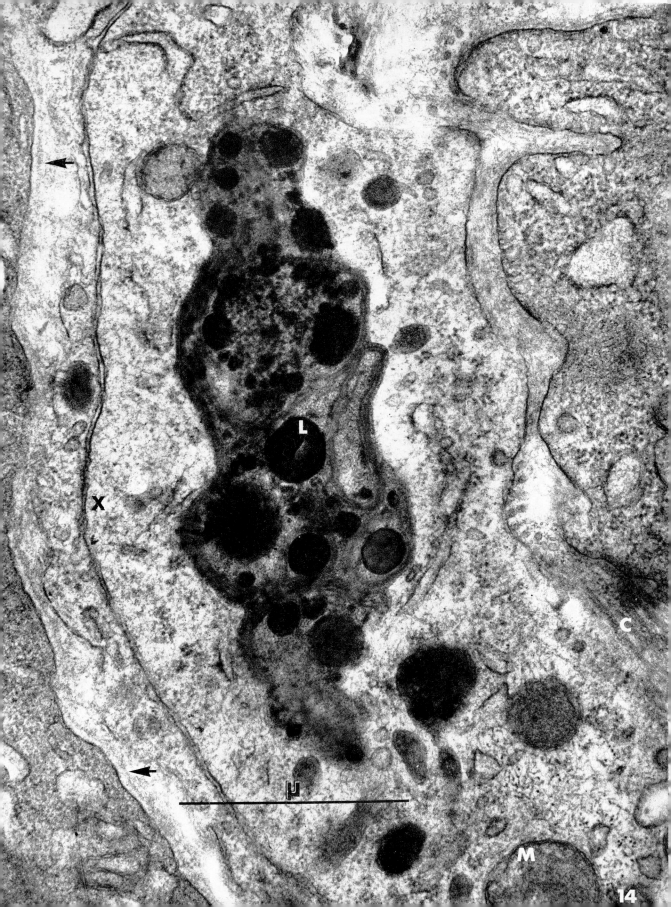

14

PLATE 15a. NUCLEAR PORE AND NUCLEAR ENVELOPE

This is a transverse section of a nuclear pore seen at high magnification. The outer and inner nuclear membranes which enclose the perinuclear cisterna come together at the pore. A small flange projects into the perinuclear cisterna at the point arrowed. The diffuse dense annulus, forming a collar around the pore, can be seen projecting into the cytoplasm. A diaphragm bridges the pore and separates cytoplasmic ground substance above from nuclear contents below. In the cytoplasm, several secretion granules surrounded by distinct limiting membranes can be seen. The contact surfaces of two cells cross the plate, but the cell membranes are not clearly seen since they have been sectioned obliquely. In the nucleus, a patch of chromatin close to the inner nuclear membrane shows the typical dense granular appearance, with no highly organised fine structure.

TISSUE Rat anterior pituitary. Osmium fixation, uranium staining.

MAGNIFICATION 172,000 ×.

REFER TO Plates 1, 12a, 16, 24.
        Pages 15, 19, 43.

PLATE 15b. NUCLEAR PORES

This is a tangential section of the nuclear envelope which gives a surface view of the nuclear pore at high magnification. The chromatin in this case is apparently arranged in strings of dense granules. Two pores, seen in face view on account of the tangential section, are present at the interface between nucleus and cytoplasm. In each case the annulus surrounding the pore has been cut across, showing a suggestion of its subunit construction. In the centre of each pore a small spot of increased density is present. In the cytoplasm nearby ribosomes, each approximately 150 Å in diameter, help to indicate the scale of the magnification.

TISSUE Rat cerebellum. Glutaraldehyde fixation, uranium staining.

MAGNIFICATION 110,000 ×.

REFER TO Plates 1, 12a, 16, 24.
        Pages 15, 19, 91.

PLATE 15c. NUCLEAR PORES

Three pores in the nuclear envelope, again sectioned tangentially, are seen between nucleus and cytoplasm. In this case the pores appear to have been sectioned at or near the level of the flange, rather than through the annulus. This plate and Plate 15b are examples of the value of tangential sections in the interpretation of the structure of parts of the cell.

TISSUE Rat cerebellum. Glutaraldehyde fixation, uranium staining.

MAGNIFICATION 110,000 ×.

REFER TO Plates 1, 12a, 16, 24.
        Pages 15, 19, 91.

A	Annulus
C	Chromatin
CM	Obliquely sectioned cell membranes
G	Granules of secretion with limiting membranes
INM	Inner nuclear membrane
M	Mitochondrion
N	Nucleus
NP	Nuclear pore
ONM	Outer nuclear membrane
P	Pore
R	Ribosomes
Y	Perinuclear cisterna
Arrow	Indicates a flange projecting into the perinuclear cisterna at the level of the diaphragm of the nuclear pore.

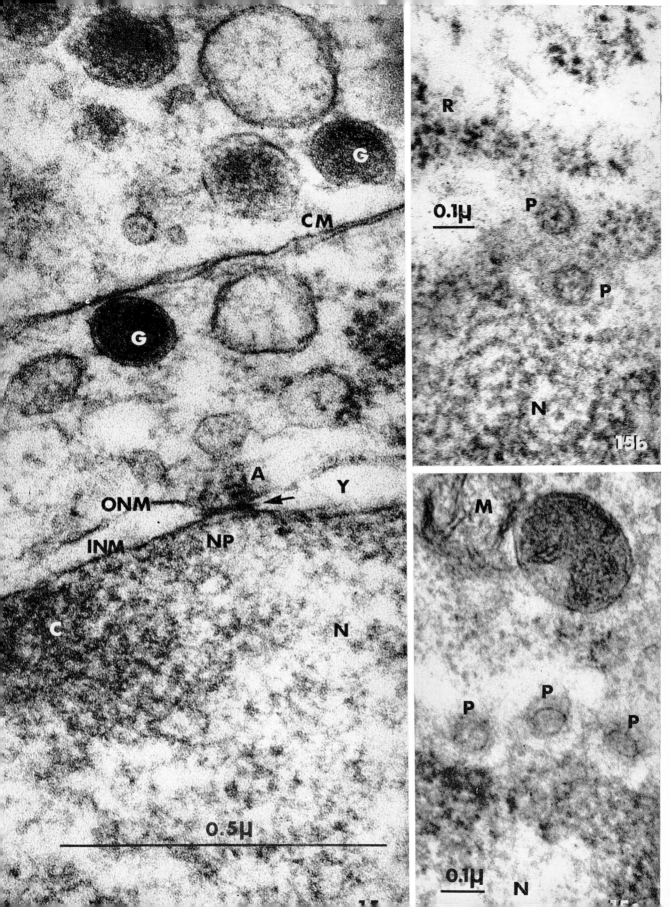

PLATE 16. NUCLEUS, NUCLEOLUS AND NUCLEAR ENVELOPE

This is a high magnification micrograph of the perinuclear cisterna, Y, which is bounded by the outer and inner nuclear membranes. The cytoplasm lies to the left of the plate, the nucleus to the right. There are two nuclear pores, at which the outer and inner nuclear membranes become continuous. Patches of chromatin lie on the inner aspect of the nuclear membrane, leaving clear channels of pale nuclear ground substance in contact with the inner aspect of the nuclear pores. The nucleolus is seen within the nucleus, easily distinguishable from the nuclear substance although not limited by a membrane. Dense granules of ribosomal dimensions can at times be distinguished within the coarsely coiled nucleolonema.

In the adjacent cytoplasm a number of components of cell structure are present. The mitochondrion indicated displays an apparent wide discontinuity, probably accounted for by the plane of section. A small portion of the Golgi apparatus is seen, with an adjacent secretion granule surrounded by a membrane.

C	Chromatin
G	Golgi apparatus
M	Mitochondrion
N	Nucleus
Nn	Nucleolonema component of nucleolus
NP	Nuclear pore
SG	Secretion granule
Y	Perinuclear cisterna

TISSUE Rat anterior pituitary. Osmium fixation, uranium staining.

MAGNIFICATION 128,000 ×.

REFER TO Plates 1, 12a, 15a, 15b, 15c.
        Pages 15, 18, 19, 91.

H

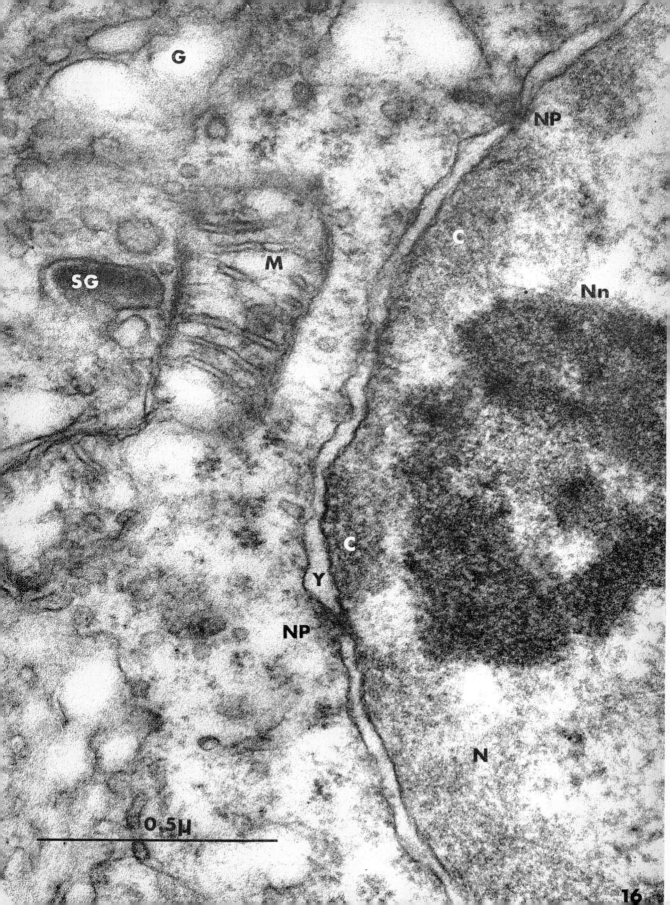

16

PLATE 17a. PERMEABILITY: CAPILLARY

In this plate, at moderate magnification, a capillary is shown adjacent to a cardiac muscle cell. The capillary, occupying the lower half of the micrograph, contains a red blood corpuscle and has in its lumen a flocculent precipitate of plasma protein. The rather thick endothelium of this vessel shows two specialisations, an adhesion point where the endothelial cells are joined together, and numerous micropinocytotic vesicles, some still in continuity with the lumen or the extracellular space, some unconnected with the cell surface, lying free within the endothelial cell cytoplasm. The flask-shaped vesicles may form at the cell surface, become filled with extracellular fluid and then pinch off from the surface membrane, passing across the cell as spherical vesicles. It is not known to what extent this forms a pathway for fluid transfer across capillary endothelium. At the right of the field a small portion of the endothelial cell nucleus is seen, surrounded by its nuclear envelope.

Immediately external to the capillary endothelium is its basal lamina or lamina densa, which separates it from the surrounding connective tissue space. The adjacent muscle cell also cut in transverse section is surrounded by its own basal lamina. Dense collagen fibres, cut in transverse section and appearing as circular profiles, are found in the intervening connective tisue space. The hexagonal filament pattern and the large mitochondria of the cardiac muscle cell are present.

TISSUE Rat heart. Glutaraldehyde fixation, lead staining.

MAGNIFICATION 47,000 ×.

REFER TO Plates 19, 21a, 21b, 30, 36.   Pages 8, 13, 49.

PLATE 17b. PERMEABILITY: CAPILLARY

This is a low magnification micrograph of a capillary containing a red blood corpuscle. The indented appearance of the surface of the corpuscle may reflect physical distortion in the capillary or may be due to irregular contraction of the corpuscle during fixation. The small pale island within the corpuscle is part of this indentation which is connected to the exterior outwith the plane of section. The endothelium is free from pores, but micropinocytotic vesicles and a poorly defined adhesion point can be seen. Notice the close relationship between the capillary basal lamina and the adjacent collagen fibres in the surrounding connective tissue.

TISSUE Rat heart. Glutaraldehyde fixation, uranium staining.

MAGNIFICATION 24,000 ×.

REFER TO Plates 19, 21a, 21b, 30, 36.   Pages 8, 13, 49, 56.

PLATE 17c. PERMEABILITY: CAPILLARY

This is a portion of the endothelial wall of a capillary in the small intestinal lamina propria, in which a red blood corpuscle is present. In this vessel the endothelium is thin and fenestrated. There are several pores, each bridged by a diaphragm which is apparently thinner than the endothelial cell surface membrane. The diffuse surrounding basal lamina is seen with its close relationship to the adjacent collagen fibres.

TISSUE Human intestine. Osmium fixation, PTA staining.

MAGNIFICATION 31,000 ×.

REFER TO Plates 18, 36, 39, 40.   Pages 13, 50, 56.

BL	Basal lamina or lamina densa
CT	Connective tissue with dense profiles of collagen fibres
F	Myofilaments in hexagonal arrangement
L	Lumen of capillary containing precipitated plasma proteins
M	Mitochondrion
N	Nucleus of endothelial cell
P	Endothelial pores
PV	Micropinocytotic vesicles
RBC	Red blood corpuscle
Z	Z line of sarcomere cut obliquely
Arrow	Indicates the point of junction between two endothelial cells at which a contact specialisation can be seen.

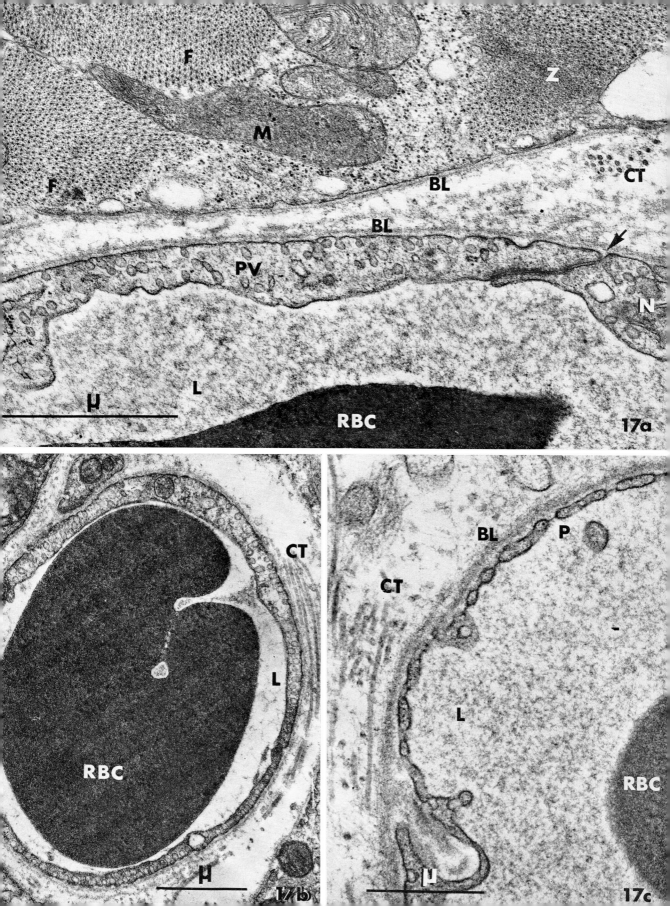

PLATE 18. PERMEABILITY: ENDOTHELIAL PORE

This micrograph shows the structure of a single capillary endothelial pore at high magnification. Compare this plate with Plate 17c. The capillary lumen lies on the right of the field, the connective tissue space on the left, while the endothelial cell crosses the micrograph vertically. The trilaminar structure of the capillary endothelial cell membrane is arrowed at one point, but no similar membrane structure is seen across the capillary pore, the diaphragm of which is more diffuse than a normal membrane. The precise nature of this diaphragm is not clear, but it is reasonable to assume that it presents an area with permeability properties which differ from those of the endothelial cell cytoplasm, which is about 400 Å thick at its thinnest point.

The capillary basal lamina or lamina densa is separated from the endothelial cell by a distinct space several hundred Ångstroms in width. A few diffuse strands of material similar in appearance to the basal lamina appear to cross this space, and are perhaps attached to the endothelial cell membrane, forming an ill-defined external coat. The basal lamina itself has no clear ultrastructural pattern, being composed of diffuse fibrillar or granular material in loose association. Even this apparent structure may represent precipitation effects, preserving, in apparently permanent form, short term associations between protein molecules which would be rapidly changing in life.

BL	Basal lamina
C	Connective tissue space
E	Endothelial cell
L	Lumen of capillary
P	Capillary pore
Arrow	Indicates area where trilaminar structure is just visible in the endothelial cell membrane.

TISSUE Rat pituitary. Osmium fixation, uranium staining.

MAGNIFICATION 410,000 ×.

REFER TO Plates 6, 17c, 20, 39, 40.
Pages 13, 50, 56.

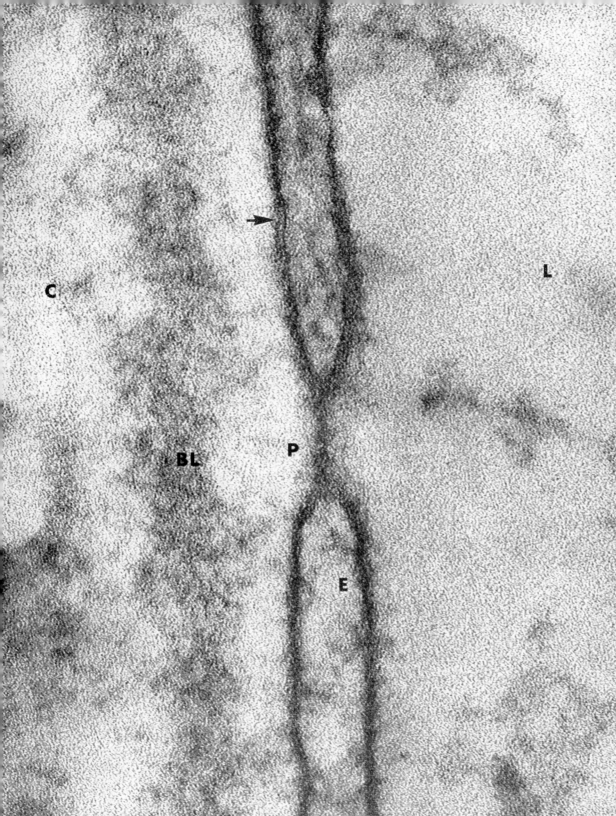

0.1 μ

18

PLATE 19. PERMEABILITY: LUNG

This is a low magnification micrograph of lung. The pale areas are the air spaces or alveoli, which are lined by a thin continuous layer of epithelium. In the upper right hand corner of the plate a great alveolar cell is seen. These cells are part of the alveolar lining epithelium, but are more cubical in shape and are secretory cells, engaged in the production of surface-acting material to reduce surface tension in the lungs. These cells have characteristic exocrine secretory inclusions and dense cytoplasm, with a few surface microvilli which indicate an epithelial nature. The thin alveolar lining cells are specialised to allow the passage of respiratory gases.

Within the alveolar wall are the pulmonary capillaries in which both red blood corpuscles and white cells can be seen. The thin continuous capillary endothelium is free from specialisations over considerable areas but shows evidence of the presence of micropinocytosis vesicles at the area indicated by En.

The epithelium and endothelium are separated only by their fused basal laminae over much of the area along which they are applied to each other. In these cases, where the epithelial and endothelial cells are thin, the total thickness of the barrier between blood and air in the lung is in the order of 1000 Å. In other places, however, a narrow connective tissue space is seen within the alveolar wall, occupied to a considerable extent by processes of pulmonary macrophages with a delicate support of connective tissue fibres.

The complex cell outlines seen at the areas marked X present a problem of interpretation. These areas represent oblique sections through a connective tissue space. The endothelium of the capillary, projecting into the lumen in an unusual configuration, has been cut obliquely at Y causing a confusing image. The slight scoring across the red blood corpuscle, typically dense and homogeneous in appearance, is a sectioning artefact, commonly seen in this situation. The flocculent appearance of the plasma within the vessels contrasts with the red corpuscles. The pale haloes surrounding the corpuscles may be partly due to shrinkage and partly due to disturbance of the tissue during processing leading to displacement of the corpuscles within their suspending plasma.

AC	Great alveolar cell
C	Capillary lumen
En	Endothelium showing pinocytotic vesicles
Ep	Attenuated alveolar lining epithelium
G	Secretion granules
PA	Pulmonary alveolar air space
PM	Pulmonary macrophage in connective tissue
RBC	Red blood corpuscle
W	White blood cell
X	Connective tissue space
Y	Endothelium, oblique section

TISSUE Mouse lung. Glutaraldehyde fixation, uranium staining.

MAGNIFICATION 12,000 ×.

REFER TO Plates 14, 17, 20.
Pages 12, 51, 54, 91.

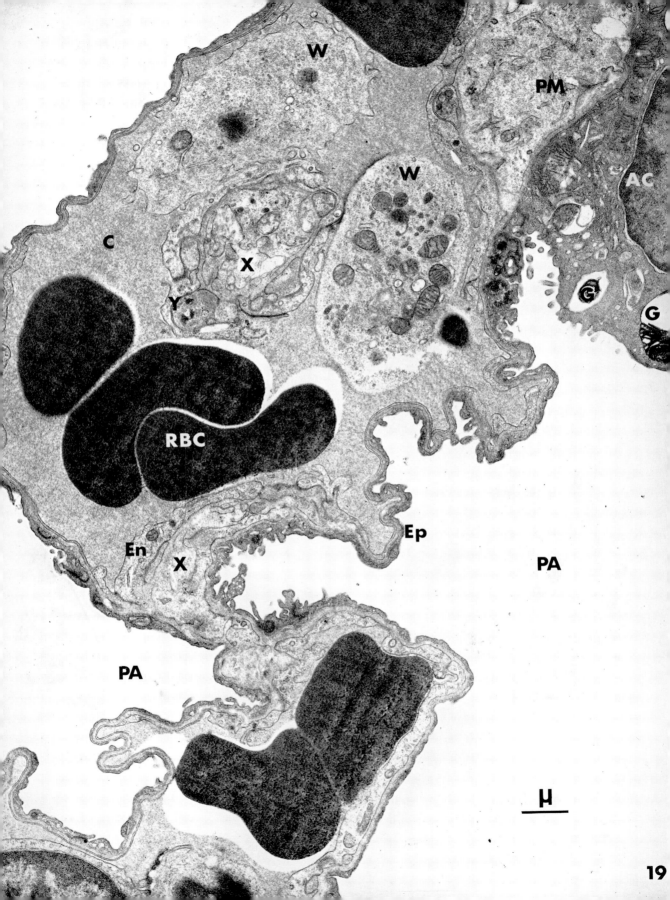

PLATE 20. PERMEABILITY: LUNG

This high magnification micrograph shows in detail the narrow interface between blood and air. The airspace of the pulmonary alveolus is to the right of the plate, the dense area on the left is part of a red blood corpuscle. The red corpuscle comes very close to the endothelial wall of the capillary. The narrow space between the surface membrane of the red corpuscle and the cell membrane of the endothelial cell is indicated by arrows. This space is occupied by plasma. The endothelial wall is less than 200 Å thick at points.

The basal laminae of endothelium and epithelium are fused in this region, no true connective tissue space remaining between them. The relatively featureless appearance of the material composing the basal lamina is typical of this structure.

The alveolar lining is formed by the alveolar epithelial cell, which is itself only a few hundred Ångstroms in thickness and is relatively featureless. In this area of the lung, the total width of the tissue separating blood from air is in places as little as 1000 Å, thus allowing rapid passage of carbon dioxide and oxygen across the barrier in accordance with the physical constants which determine diffusion rates. In view of the surprising delicacy of this barrier, with the almost total absence of supporting tissues over large areas, it may seem surprising that the alveolar walls can resist the mechanical stresses of respiration.

BL	Basal lamina or lamina densa
E	Endothelial cell
Epi	Alveolar epithelium
PA	Pulmonary alveolus, air space
RBC	Red blood corpuscle
Arrows	Indicate the narrow gap between red corpuscle and endothelial cell.

TISSUE Mouse lung. Glutaraldehyde fixation, uranium staining.

MAGNIFICATION 210,000 ×.

REFER TO Plates 6, 18, 19, 39, 40.
Pages 12, 51, 54.

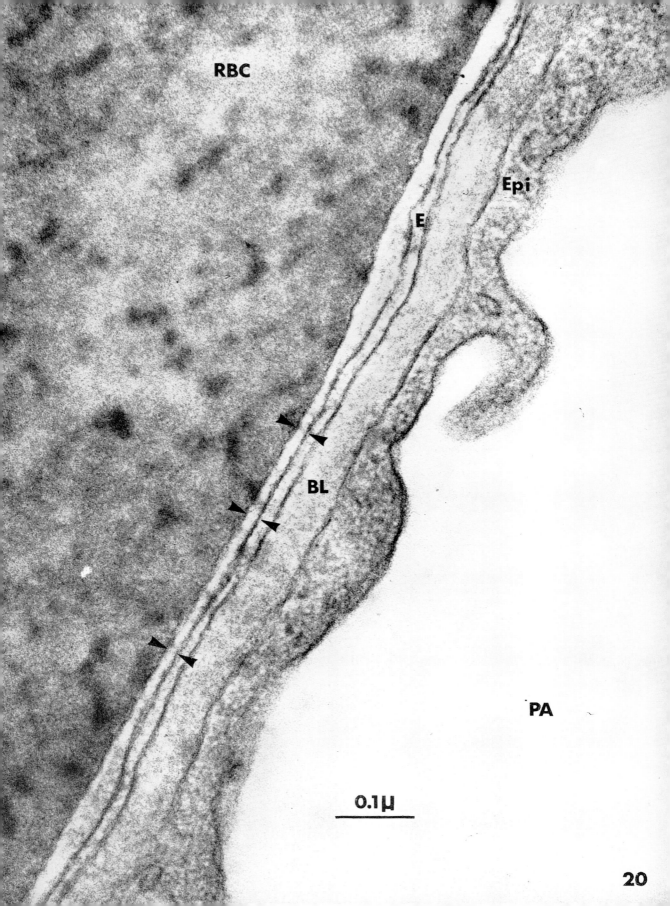

RBC

Epi

E

BL

PA

0.1μ

20

PLATE 21a. CONNECTIVE TISSUE

This is a low magnification micrograph of an area of connective tissue between the semini-ferous tubules of the testis. The cell with dense inclusions in the upper right hand corner of the plate is an interstitial cell of the testis, which secretes steroid hormones. The dense inclusions, perhaps lipid in nature, may be stored aggregates of secretion precursor. Some of the features of a steroid secreting cell are seen, such as the prominent smooth endoplasmic reticulum and the mitochondria with their apparently tubular cristae.

Parts of several other cells are seen, but they are widely separated, unlike the cells of an epithelium. The majority of cells are fibroblasts. In many cases only part of a cell is present, often without a nucleus being seen. These represent the processes of cells which lie mainly outside the plane of section. In some of them there is apparent membrane damage suggesting that there has been trauma to the tissue during the selection of the sample. The absence of the membrane and the partial dispersal of the cell components indicates damage of this type at X.

The separation of the cells in this plate is typical of loose connective tissue. There are relatively few collagen fibres present in this micrograph, although several can be seen at the point arrowed. A capillary, a common finding in connective tissue, can be seen in the lower part of the field.

C	Capillary
F	Fibroblast
G	Granule of interstitial cell
M	Mitochondrion
N	Nucleus
SER	Smooth endoplasmic reticulum
X	Damaged cell
Arrow	Indicates several isolated collagen fibres.

TISSUE Mouse testis. Osmium fixed, uranium staining.

MAGNIFICATION 14,000 ×.

REFER TO Plates 14, 24, 34a.
Pages 44, 49, 56, 90.

PLATE 21b. COLLAGEN

This is a high magnification micrograph of a single collagen fibre. The arrows point to re-curring points in the characteristic repeating pattern of collagen. Within each repeating period, measuring about 640 Å in length, a number of discrete bands can be detected, in this case about 7 or 8. This repeating banded pattern is believed to arise from the ordered parallel stacking of the elongated tropocollagen molecules which compose the collagen fibre. Collagen is the most distinctive component of connective tissue and its cross banding is a guide to magnification if there is no scale provided in a micrograph.

Arrows   Indicate successive corresponding points in the pattern of repeating cross banding which characterises the collagen fibre.

TISSUE Mouse epididymis. Glutaraldehyde fixation, uranium staining.

MAGNIFICATION 170,000 ×.

REFER TO Plates 6, 17a, 35.
Pages 56, 90.

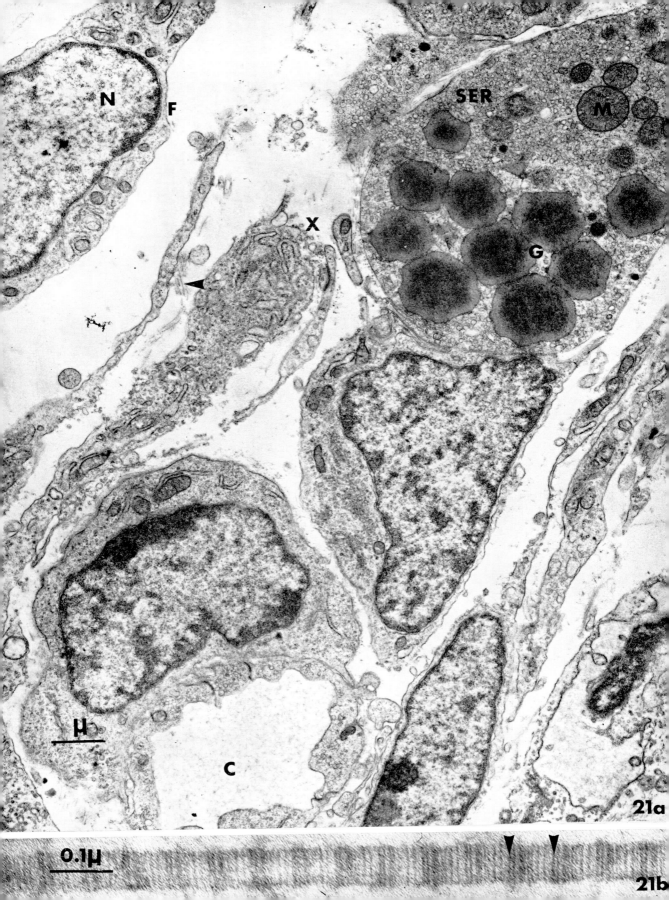

N F

SER M

X

G

◄

μ

C

0.1μ

21a

21b

PLATE 22. EXOCRINE SECRETION

This is a low magnification micrograph of cells which line the stomach of the chick embryo, producing a form of mucus secretion. The cells project into the lumen and their bulging apical surfaces are covered with irregular microvilli. Adjacent cells are joined by junctional complexes and desmosomes. These features indicate that these are epithelial cells. There is a small intercellular space crossed by projections from adjacent cells.

There are numerous granules present in the cytoplasm, some of which are apparently being released at the apical surface of the cell into the lumen of the stomach. The granule morphology is variable, some being denser and perhaps more mature than others. Dehydration of the granule contents may be responsible for the increased density shown by some of them. Once the granules are discharged from the cell into the lumen they become swollen and flocculent in appearance and rapidly become dispersed. The pale vacuoles at the apex of the cell are probably evidence of secretion granules not yet fully discharged. The membrane surrounding the granule at V has probably fused with the cell surface outside the plane of section and the contents of the granule, altered by uptake of fluid from the lumen, have not yet become fully dispersed.

The cytoplasmic features associated with this type of exocrine secretion are the complex lamellar Golgi apparatus with its surrounding vesicles and several closely associated newly formed secretion granules. The granular endoplasmic reticulum is more scattered than in the plasma cell, but is still a prominent feature of the cell, although more marked in the base of the cell which is not shown in this plate. Mitochondria are present in moderate numbers. The nucleus of one cell, on the left of the field, is very indented and irregular in outline. A portion of another nucleus lies at the foot of the micrograph.

D	Desmosome
G	Granules of secretion, some dispersing in the lumen, some enclosed within the cytoplasm
GER	Granular endoplasmic reticulum
Go	Golgi apparatus
JC	Junctional complex
L	Lumen
M	Mitochondrion
MV	Microvilli
N	Nucleus
V	Vacuole caused by discharge of secretion

TISSUE Chick embryo gizzard. Osmium fixation, lead staining.

MAGNIFICATION 29,000 ×.

REFER TO Plates 7, 9a, 9b, 23.
          Pages 22, 25, 37, 90.

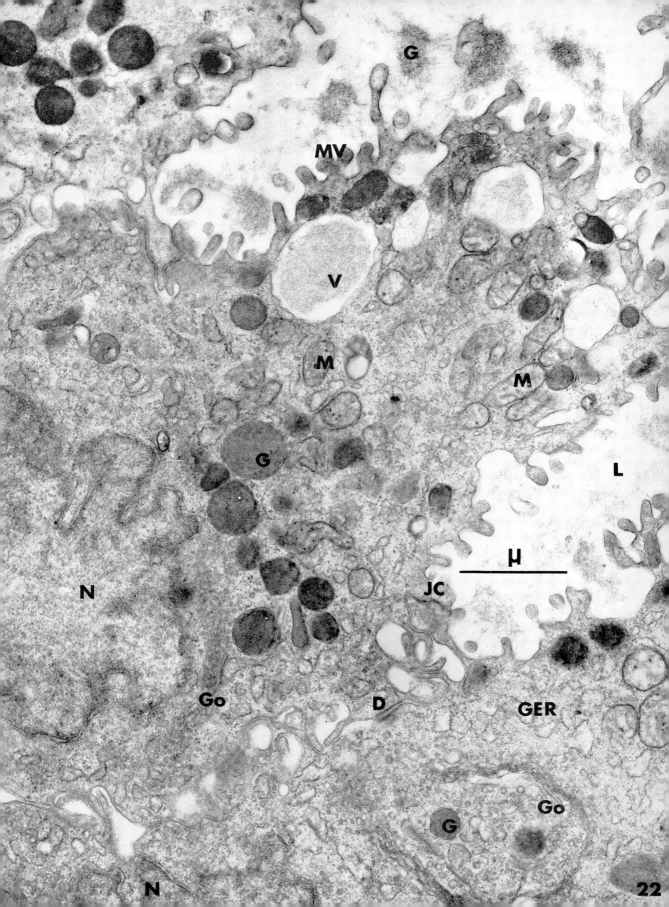

22

PLATE 23. EXOCRINE SECRETION

Parts of two epithelial cells are seen at high magnification in this plate. They are gland cells from the seminal vesicle, which have a protein-secretory function. The gland lumen lies to the right of the field and contains an amorphous mass of secretion. Material of similar density is found in large cytoplasmic vacuoles, one of which is apparently on the point of rupturing at the cell surface and discharging its contents. The lowest granule in the field is attached to the margin of its containing vacuole.

The apical surfaces of the cells carry a few stubby microvilli, irregularly arranged. The surface membrane shows a trilaminar structure at several points including those arrowed. This appearance must be distinguished from the close apposition of the two adjacent lateral cell membranes at the contact surfaces of the cells indicated by CM. Each of the dense lines seen here represents a separate membrane, the pale intervening gap being the narrow extra-cellular space typical of epithelial tissues. At points such as at X due to irregularity of the cell membranes, the clear-cut appearance of the contact surfaces becomes diffuse on account of oblique sectioning. A membrane appears completely sharp only when it is sectioned at right angles to the plane which it forms.

The cytoplasm is dense, partly on account of the many ribosomes which it contains, often attached to the surfaces of the cisternae of the granular endoplasmic reticulum which fill most of the cell. The Golgi apparatus is prominent in these exocrine secretory cells although only a small portion of it can be seen in this micrograph.

CM	Cell membranes in contact on lateral cell surfaces
G	Golgi apparatus
GER	Granular endoplasmic reticulum
S	Secretion lying in lumen
SG	Secretion granule in vacuole
X	Diffuse area of membrane probably due to oblique sectioning effect
Arrows	Indicate trilaminar structure at points on the surface membrane of the cell.

TISSUE Mouse seminal vesicle. Osmium fixation, lead staining.

MAGNIFICATION 54,000 ×.

REFER TO Plates 2, 7, 22.
Pages 5, 22, 25, 37, 91.

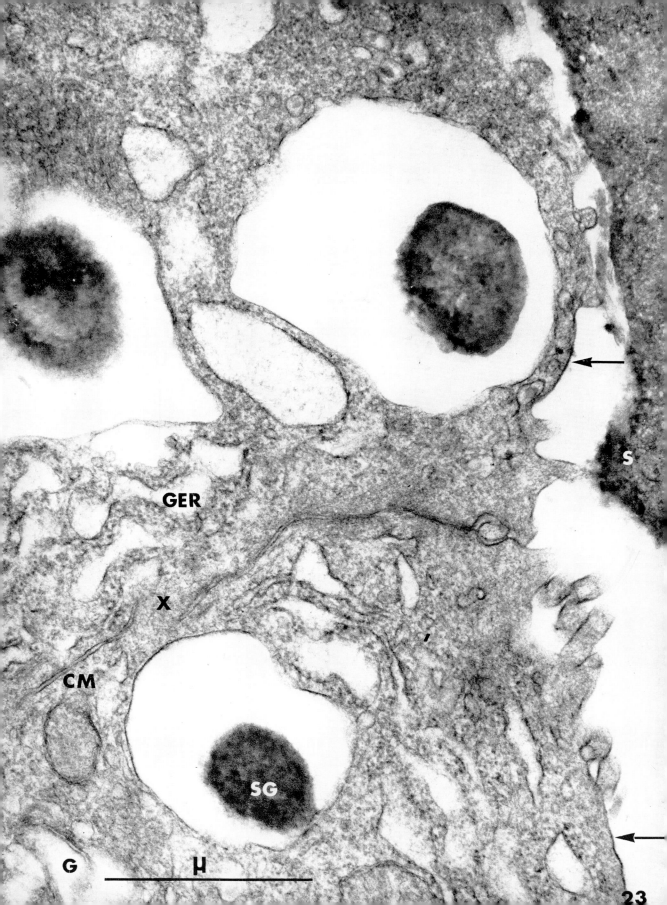

GER

X

CM

SG

G   μ

S

23

PLATE 24. ENDOCRINE SECRETION

This is a low magnification micrograph of the anterior pituitary of the rat. The main feature of the cells is the presence of numerous small dense cytoplasmic granules, some of which show their limiting membranes. This fine granular appearance is a common feature of several endocrine glands, although details of cell structure vary greatly. The base of the cell which occupies most of the centre of the field lies to the right of the micrograph and is close to the basal lamina which separates the cells from the connective tissue space. There is no predominantly basal accumulation of granules in this cell, such as may sometimes be seen in other endocrine cells.

The cytoplasm is not highly organised. The mitochondria are small and are difficult to distinguish clearly. The few cisternae of the endoplasmic reticulum are isolated from each other and appear vesicular in profile. A Golgi apparatus lies close to the nucleus. The nucleus itself, seen in the centre of the plate, is a guide to the relatively low magnification. Nuclear pores can just be distinguished on careful examination of the nuclear envelope. Ill-defined dense patches of chromatin are located peripherally, separated by paler nuclear ground substance which forms channels in contact with the inner aspect of the pores.

Parts of adjacent cells lie in close contact, but the cellular arrangement is slightly irregular, giving rise to misleading 'islands' of cytoplasm marked X belonging to cells which lie mainly out of the plane of section. Although a few intercellular spaces are seen, this part of the field is not of connective tissue nature.

On the right of the micrograph, on the connective tissue side of the basal lamina, there is part of a capillary with a pale lumen and thin endothelial lining in which endothelial pores or fenestrations are present. The capillary is surrounded by a discrete close-fitting basal lamina. The hormone discharged by the endocrine cells shown here must cross two basal laminae and the intervening connective tissue space, as well as the wall of the capillary endothelium, before reaching the blood stream. The thickness of this barrier, however, is often much less than one micron. The presence of endothelial fenestrations may reflect permeability properties which are different from those seen in other capillaries.

BL	Basal lamina or lamina densa
C	Connective tissue space
E	Endothelial cell
ER	Granular endoplasmic reticulum
G	Golgi apparatus
ICS	Intercellular space
M	Mitochondrion
N	Nucleus
P	Endothelial pores
X	Processes of cells which lie mainly out of the plane of section
Arrows	Indicate nuclear pores and corresponding gaps between chromatin patches in the nucleus.

TISSUE Rat pituitary. Osmium fixation, lead staining.

MAGNIFICATION 18,000 ×.

REFER TO Plates 6, 15a, 16, 17c, 18.
        Pages 12, 15, 43, 49.

I

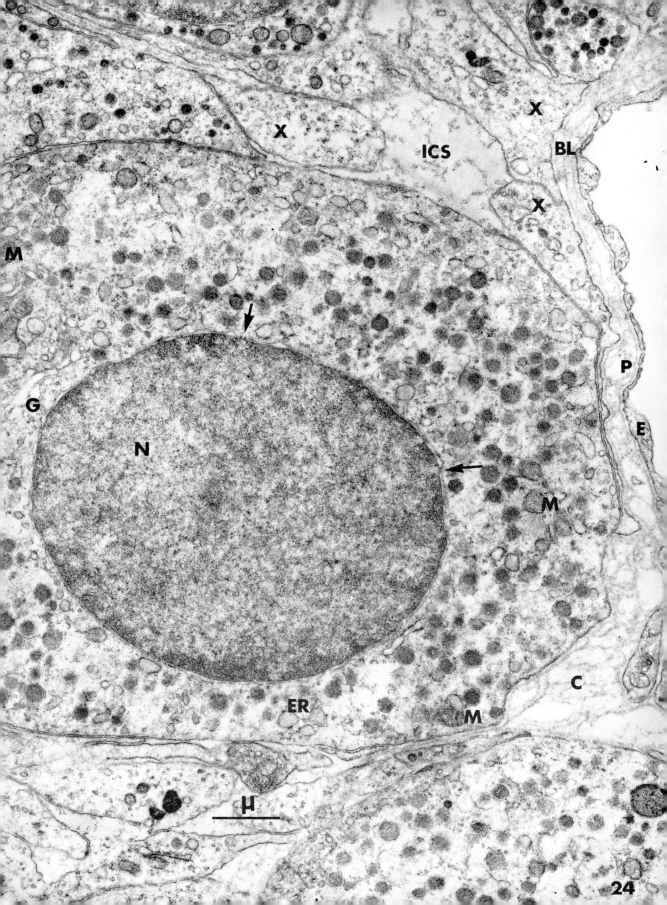

PLATE 25a. ABSORPTION AND DIGESTION: SMALL INTESTINE

This plate shows the apical parts of several intestinal absorptive epithelial cells from normal human small intestine, at low magnification. The material was obtained by intestinal biopsy. The columnar nature of the epithelial cells can be readily seen, with only a narrow 150 Å gap being present between adjacent cells. At the contact surfaces of the epithelial cells the apically placed junctional complexes are seen, while lower down there are interdigitations between contact surfaces and desmosomes which are just visible at this low magnification.

The 'striated border' of the intestinal cell is seen at the cell apex as a discrete horizontal band in this plate. The microvilli which form this border are closely packed and are coated with a surface layer, the cell coat or intestinal fuzzy coat, which is indicated by an arrow. The microvilli are cut slightly obliquely, so that the full length of each individual process does not appear in the section. The terminal web lies below the microvilli. Its function may be to stiffen the cell apex. Filaments pass from it to the cores of the microvilli. The terminal web is associated also with the junctional complex situated at the apical parts of the contact surfaces. The mitochondria of the intestinal cells are dense in appearance and are quite numerous. Dense structures which may be lysosomes appear in small numbers in the cell apex.

TISSUE Human small intestine. Osmium fixation, lead staining.

MAGNIFICATION 8000×.

REFER TO Plates 4a, 4b, 5, 25b.
        Pages 7, 9, 11, 44.

PLATE 25b. MALABSORPTION: SMALL INTESTINE

This plate shows the apical half of several intestinal absorptive epithelial cells from a case of malabsorption syndrome due to gluten sensitivity. The material was obtained by intestinal biopsy. The cells are abnormal in several ways. The microvilli are shorter than the normal seen above and are less regularly arranged and less closely packed, although the surface fuzzy coat is well seen. Abnormal large vacuoles are seen in the cytoplasm, while dense lysosome-like structures are now more numerous than in the normal cell. There is perhaps a slight reduction in the numbers of mitochondria present. At the points marked X, parts of a small lymphocyte are seen, the cell apparently migrating through the epithelium by pushing between the epithelial cells. Although small lymphoid cells of this type are seen in normal small intestine, their numbers are increased in disease of this kind.

TISSUE Human small intestine. Malabsorption syndrome. Osmium fixation, lead staining.

MAGNIFICATION 8000×.

REFER TO Plates 4a, 4b, 5, 14, 25a.
        Pages 7, 9, 11, 44, 47.

CM	Cell membranes at contact surfaces
D	Desmosome
JC	Junctional complex
L	Lysosome-like structure
Lu	Lumen of intestine
M	Mitochondrion
MV	Microvillous border
N	Nucleus
TW	Terminal web
V	Vacuoles
X	Cytoplasm of migratory lymphoid cell
Arrow	Indicates fuzzy coat covering intestinal microvilli.

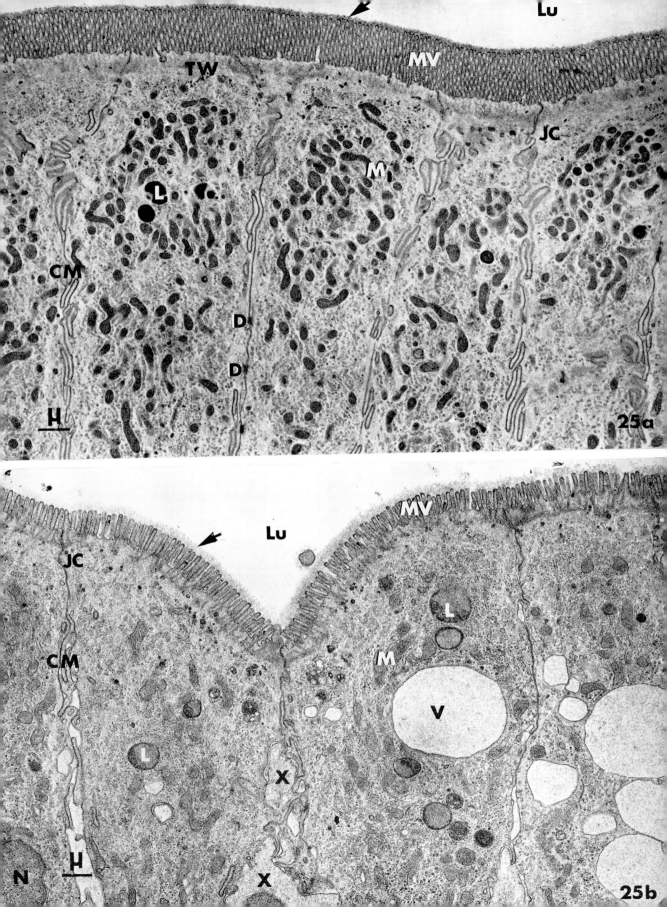

Lu

MV

TW

JC

M

L

CM

D

D

H

25a

JC

Lu

MV

CM

L

M

L

V

X

N

H

X

25b

PLATE 26. CONTRACTION: SKELETAL MUSCLE

This micrograph, at moderate magnification, shows a small part of a single skeletal muscle cell cut in longitudinal section. The long axis of the muscle cell runs from left to right. The surface membrane of the cell is not seen in this plate, which shows only a few centrally placed myofibrils. These myofibrils, separated by sarcoplasm, cross the plate horizontally. The alignment of the sarcomeres of each myofibril gives rise to the striated appearance of the muscle cell. Each myofibril shows the whole of a single sarcomere, extending from one Z line to the next, along with parts of two other adjacent sarcomeres.

In the dense A band the thick myosin filaments are present, while in the narrower I band only thin actin filaments are found. Over part of each end of the A band, indicated by arrows at O, the thick and thin filaments overlap and interdigitate. Around the midpoint of the A band there are only thick filaments without overlapping thin filaments. This is the H zone. Each thick filament is thickened or beaded at its midpoint, giving rise to the M line, the thickenings of adjacent filaments corresponding in position to each other. The M line therefore bisects both the H zone and the A band. In the I band the thin filaments are joined together by the Z line, which bisects the band. The Z line marks the boundary between sarcomeres, each of which consists therefore of an A band flanked by two half I bands.

When the muscle contracts, the thin filaments are thought to slide between the thick, causing shortening of each sarcomere. The ends of the thin filaments come closer to the M line, leading to narrowing of the H zone, while the I band also decreases in width. During contraction there is no significant physical shortening of the individual thick and thin filaments. There is only a shift in their position relative to each other in every sarcomere which leads to shortening of the entire myofibril.

Each myofibril is separated from its neighbour by a small volume of muscle cell cytoplasm, usually termed sarcoplasm. Within the sarcoplasm are found mitochondria which are arranged in places in a distinct segmented pattern, corresponding to that of the sarcomeres. They appear to be preferentially grouped in relation to the I band in this case. The sarcoplasmic reticulum fills the remaining space in the sarcoplasm, its interconnecting cisternal nature being best seen at SR, where it is sectioned tangentially, since it forms a sleeve around each individual myofibril. The T system along with the adjacent flanking foot processes of the longitudinal portion of the sarcoplasmic reticulum form the triads which can be seen at points close to the junctions between the A band and the I band of each sarcomere.

A	A band
H	H zone
I	I band
M	M line
Mi	Mitochondrion
O	Extent of overlap between thick and thin filaments
SR	Sarcoplasmic reticulum, showing cisternal arrangement
T	Triads with T tubules

TISSUE Rat skeletal muscle. Osmium fixation, PTA staining.

MAGNIFICATION 40,000 ×.

REFER TO Plates 8, 27, 28a, 28b, 29, 30.
         Pages 24, 60, 64.

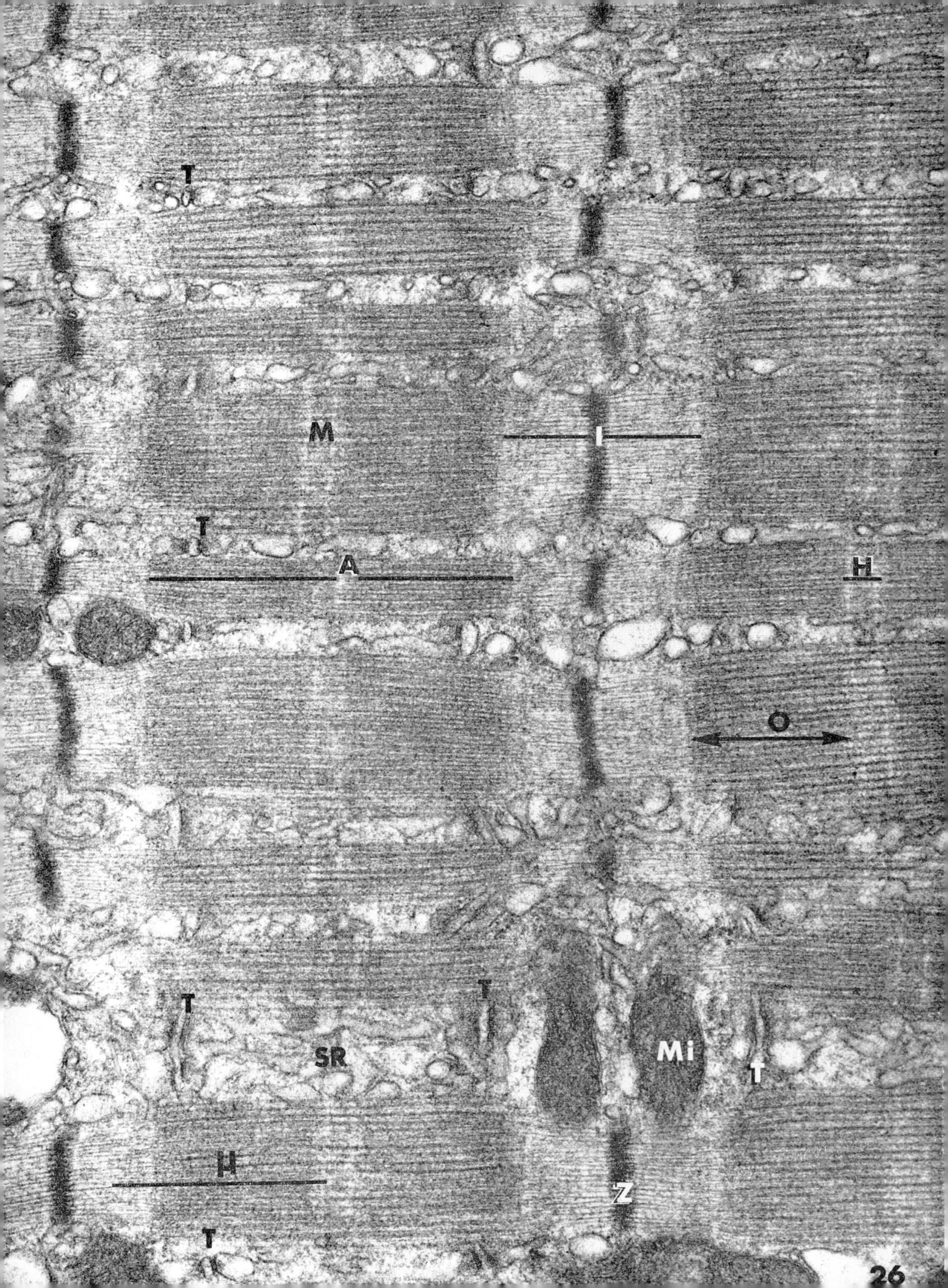
T
M
I
T
A
H
O
T
T
SR
Mi
T
H
Z
T
26

PLATE 27. CONTRACTION: SKELETAL MUSCLE

This is a high magnification micrograph of part of a skeletal muscle cell showing several parallel myofibrils crossing the plate vertically. They are separated by sarcoplasm. Part of a single sarcomere is seen. This plate should be compared directly with the foregoing plate, which shows the same features at lower magnification. Notice that the direction of the myofibrils in this plate is at right angles to that seen in Plate 26.

The Z line bisects the pale I band which contains thin actin filaments. The thick filaments occupy the A band. At the point in the A band marked by A, the overlap between thick and thin filaments is seen. In the centre of the A band lies the H zone, in which only thick filaments are present. The beadings at the midpoints of the thick filaments line up to form the M line.

The five myofibrils which cross the field almost vertically are separated by sarcoplasm in which the components of the sarcoplasmic reticulum are seen. The triads, with central T tubule and flanking foot processes, are present consistently at the region of the A-I junction, being particularly well seen at the points arrowed. The sarcomere pattern established by the arrangement of thick and thin filaments is followed by the sarcoplasmic reticulum and the mitochondria.

A	A band
H	H zone
I	I band
M	M line
Mi	Mitochondrion
SR	Sarcoplasmic reticulum
T	T tubule with flanking foot processes
Z	Z line
Arrows	Indicate the triads located close to the A-I junction.

TISSUE Rat skeletal muscle. Osmium fixation, PTA staining.

MAGNIFICATION 105,000 ×.

REFER TO Plate 26.
Pages 24, 60, 64.

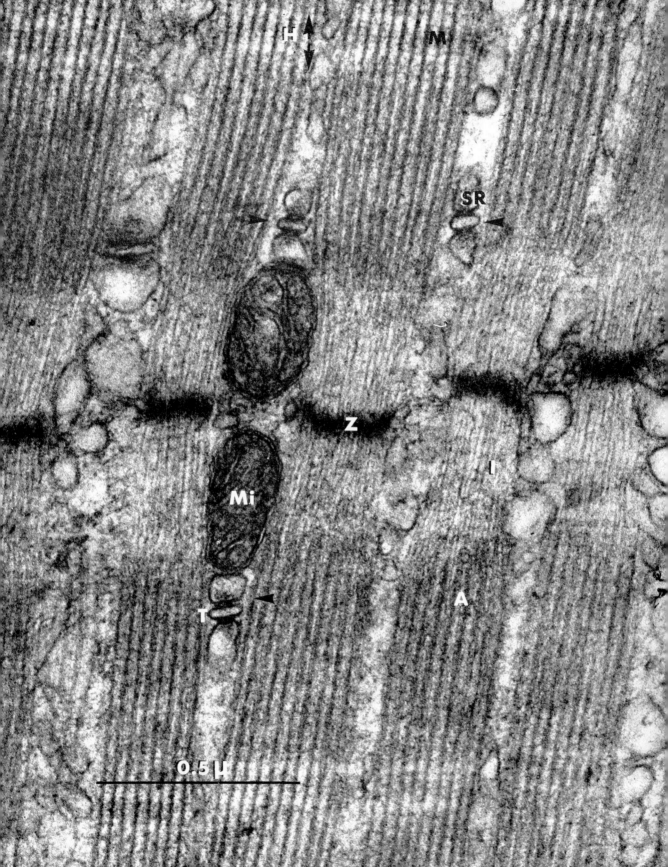

Plate 28a. Skeletal Muscle: Glycogen

This is a micrograph at moderate magnification of part of a skeletal muscle cell. The sarcomere pattern is less clearly seen than in the preceding plates, partly because the muscle is in a more contracted state. The Z lines can, however, be distinguished, and the thick A band filaments can be seen. The surface of the muscle cell is seen, with adjacent connective tissue. The external basal lamina or lamina densa of the muscle cell is closely applied to the surface of the cell and separates it from the connective tissue space. The basal lamina has no organised fine structure and blends into the connective tissue.

Within the sarcoplasm is seen an aggregation of dense glycogen particles closely associated with the cisternae of the sarcoplasmic reticulum. These granules, about 300 Å diameter, are larger than ribosomes and ferritin particles, and are readily distinguished from them in thin section.

TISSUE Rat skeletal muscle. Glutaraldehyde fixation, lead staining.

MAGNIFICATION 53,000 ×.

REFER TO Plates 7, 13a, 26.
          Pages 12, 23, 63, 67.

Plate 28b. Cardiac Muscle: Transverse Section

This is a tranverse section of part of a cardiac muscle cell, showing at high magnification the interrelationships of the overlapping thick and thin filaments. The myosin filaments appear in cross section as the larger round dense profiles, one of which is arrowed. The actin filaments are the smaller surrounding spots. A single myofibril, composed of many myofilaments, extends in this plate between the two mitochondria designated M.

Since the overlapping of thick and thin filaments occurs only in the A band the section shown here must pass through that portion of the sarcomere. If only thick filaments were seen, the plane of section would lie in the H zone: if only thin filaments were present, the plane of section would lie instead in the I band. In places a hexagonal stacking pattern of filaments can be seen. The myofibril which occupies most of this micrograph is not surrounded by a membrane. Its component myofilaments come into direct close contact with the sarcoplasm.

TISSUE Rat cardiac muscle. Glutaraldehyde fixation, uranium staining.

MAGNIFICATION 160,000 ×.

REFER TO Plates 17a, 26.
          Page 63.

A	A band
BL	Basal lamina or lamina densa
CT	Connective tissue space
G	Glycogen
M	Mitochondrion
SR	Sarcoplasmic reticulum
Z	Z line
Arrow	Indicates a thick myosin filament, surrounded by thin actin filaments.

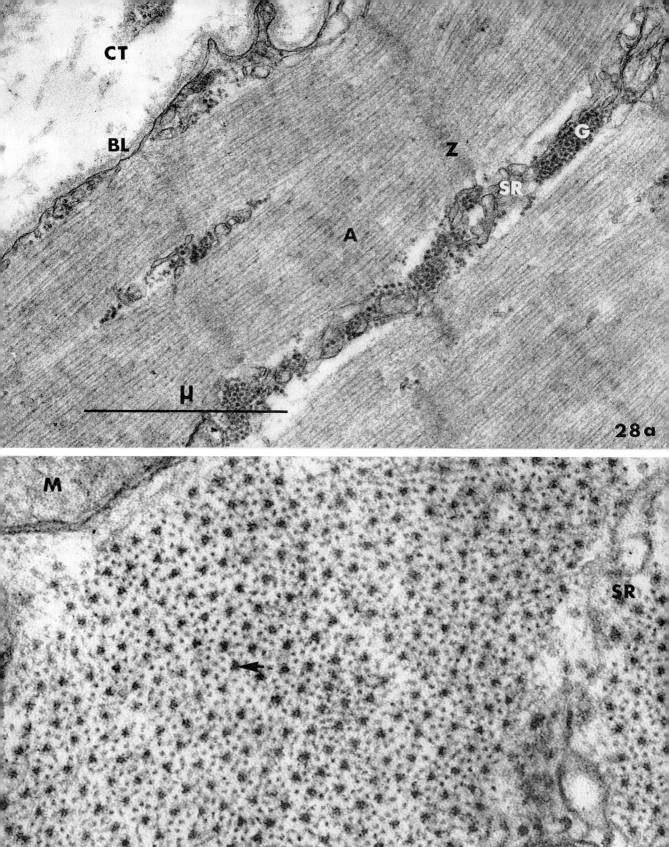

CT

BL

Z

G

SR

A

μ

28a

M

SR

M

0.1μ

28b

PLATE 29. CONTRACTION: SMOOTH MUSCLE

This is a low magnification micrograph of several smooth muscle cells. Adjacent cells are separated by a connective tissue space of variable width, in which sparse collagen fibres can be seen at a number of points reinforcing the delicate basal lamina or lamina densa which surrounds each cell individually. There is one close junction present, at which there is fusion of the membranes of adjacent cells perhaps providing a point at which activation may spread from cell to cell.

Most of the cytoplasm is occupied by filaments with a predominantly longitudinal orientation without evidence of a sarcomere pattern of repeating units. Interspersed between the fine filaments are dense bodies of uncertain length which often appear cigar-shaped in section. Thickenings on the inner aspect of the cell membrane at different points may form attachments for the myofilaments and may serve for adhesion between cells. Micropinocytotic vesicles appear at the cell surface at several points.

Few formed organelles are present in the cell. The mitochondria are relatively poorly developed and form only a small proportion of the cytoplasmic volume. Ribosomes and a small Golgi apparatus are sometimes present close to the nucleus in smooth muscle. The nucleus of one cell appears in two portions in this plate. The nucleus in smooth muscle is commonly twisted in configuration and a thin section which fails to pass through its centre may give a false impression that more than one nucleus is present in the cell. The two nuclear profiles seen in the upper right hand corner of this plate are probably the result of such a plane of section effect.

The complete contrast between the fine structure of smooth and striated muscle underlines their known differences of function.

C	Connective tissue space with collagen fibres
CJ	Close junction
D	Dense component with cigar-shaped profiles
F	Myofilaments
M	Mitochondrion
N	Nucleus
P	Micropinocytotic vesicles
R	Area of cytoplasm close to nucleus, containing ribosomes
T	Thickening on the inner surface of the cell membrane.

TISSUE Chicken gizzard, smooth muscle. Osmium fixation, lead staining.

MAGNIFICATION 16,000 ✕.

REFER TO Plates 26, 30.
        Pages 10, 12, 65.

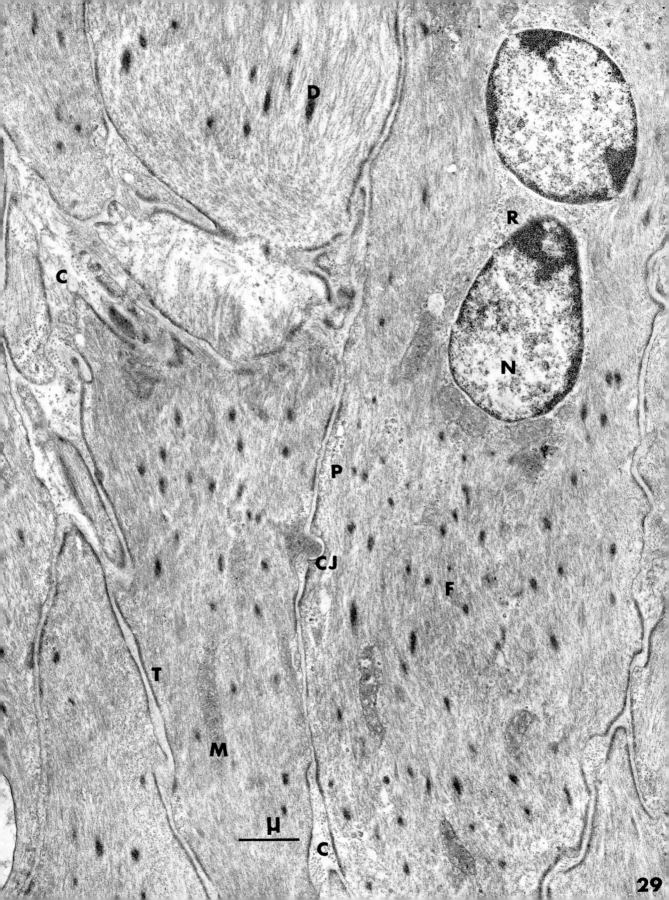

29

PLATE 30. CONTRACTION: CARDIAC MUSCLE

This micrograph shows a portion of cardiac muscle at moderate magnification. The striated pattern of the muscle is not so clearly seen as in Plates 26 and 27, but the main components of the sarcomere are distinguishable. The dense Z line is the most obvious part of the recurring pattern.

An intercalated disc passes in a zig-zag line across the field. The disc is a complex intercellular adhesion zone, and there are therefore parts of two cells in this plate. The disc represents their contact surfaces, along which different forms of contact specialisation can be seen. It can be noted that the disc replaces the Z line of the sarcomere, the density which it shows along most of its length being accounted for by the insertion of the thin filaments of the sarcomere into the cytoplasmic surfaces of the cell membranes. In this respect this part of the disc is analogous to the zonula adhaerens structure of the junctional complex and to the desmosome, with their inserted cytoplasmic fibrils. These points of the disc are prolybab concerned with mechanical adhesion, providing firm attachments against which the myofibrils can contract. The two points of the disc marked D at the upper left hand corner and lower right hand corner of the plate show areas of close junction specialisation. Such areas are believed to allow passage of excitation from cell to cell. At points marked Y the clear-cut appearance of the disc is not seen, probably because the membranes have turned at an oblique angle to the plane of section.

In the sarcoplasm, which is quite bulky in proportion to the myofibrils, the interconnecting cisternae of the sarcoplasmic reticulum can be seen, particularly when a 'surface view' of the cisternae is gained as they anastomose over the surface of a tangentially sectioned myofibril. Such an appearance is seen at the upper right hand corner of the plate. The T system is not particularly clear in this case. A droplet which is probably lipid in nature, slightly distorted perhaps by compression during preparation, is seen in the sarcoplasm. A close association between lipid droplets and mitochondria is seen in cardiac muscle. The mitochondria, large and numerous, show close-packed parallel cristae.

D	Intercalated disc
L	Lipid droplet
M	Mitochondria
SR	Sarcoplasmic reticulum, cut obliquely
Y	Oblique section of membranes of intercalated disc
Z	Z lines

TISSUE Rat cardiac muscle. Glutaraldehyde fixation, uranium staining.

MAGNIFICATION 32,000 ×.

REFER TO Plates 11, 26, 27, 29.
Pages 10, 66, 67.

PLATE 31. CARDIAC MUSCLE: INTERCALATED DISC

This plate shows a portion of the intercalated disc of cardiac muscle at high magnification. Part of a sarcomere can just be distinguished in the lower half of the plate. The Z line marks the termination of the sarcomere and as usual the contact surfaces of the cells take the place of the Z line at the intercalated disc. The main specialisations of the disc appear in this micrograph. There is an area of close junction of moderate extent, forming perhaps a region of ionic permeability between cells. The adhaerens structure, marked A, provides an attachment for the thin filaments and thus an anchorage for the mechanical pull of the myofibril.

The mitochondria are characteristically large and elaborate. The cristae are closely packed and the mitochondrial mass is considerable in proportion to the other components present. A few intramitochondrial granules are present. There is a small area of extracellular space indicated by Y, where a distinct basal lamina can be seen, applied to the muscle cell surface. This indicates that the area is at the edge of the muscle cell. Some of the mitochondria are sectioned obliquely, causing blurring of their membranes and at times a false impression of rupture. The mitochondrion arrowed has been cut in this way. Notice the trilaminar structure just visible in the mitochondrial membranes.

A	Adhaerens specialisation
CJ	Close junction or occludens specialisation
M	Mitochondrion
Y	Intercellular gap with basal lamina applied to the cell surface
Z	Z band
Arrow	Indicates obliquely sectioned mitochondrion with blurred outline.

TISSUE Rat cardiac muscle. Glutaraldehyde fixation, uranium staining.

MAGNIFICATION 88,000 ×.

REFER TO Plates 2, 3a, 26, 30.
Pages 10, 27, 66, 67.

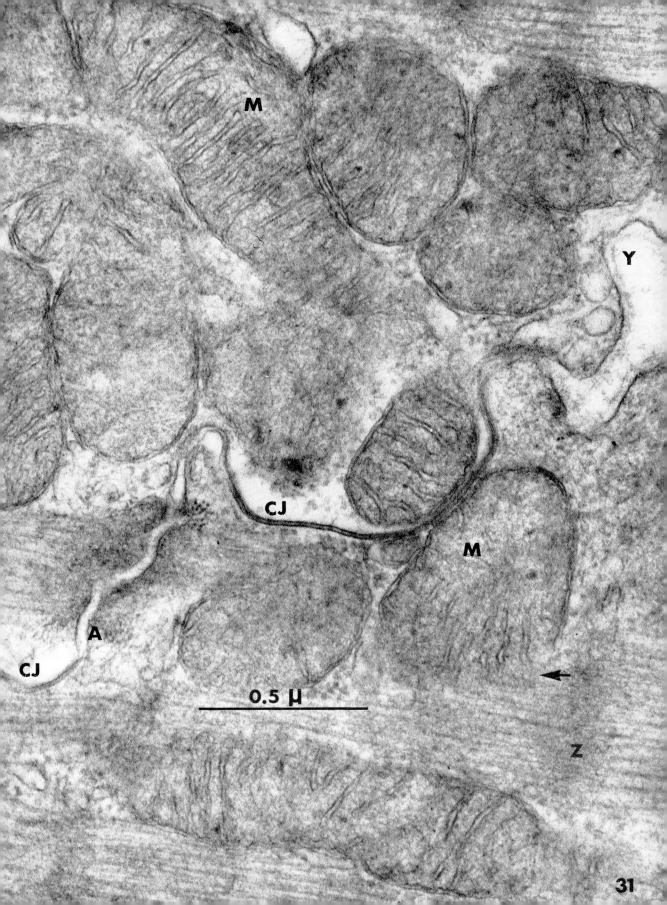

31

PLATE 32. SKELETAL MUSCLE: MOTOR END PLATE

This is a low magnification micrograph showing a motor nerve terminal entering into contact with a skeletal muscle cell at the motor end plate. The muscle cell occupies the main part of the field while the nerve terminal reaches it from the connective tissue space in the upper part of the micrograph.

The motor nerve enters a depression on the surface of the muscle cell at the point marked X-X and branches to form terminals in close contact with the infolded muscle cell surface. The small branches of the nerve, two of which are seen in this part of the end plate, are distinguished by the presence of numerous small vesicles, thought to contain the stored transmitter substance, acetyl choline, which is released from the nerve terminals on stimulation. The surface membrane of the nerve terminal is not folded or specialised in any obvious way.

The corresponding surface of the muscle cell is thrown into elaborate folds around each nerve terminal. The space between the nerve and the muscle cell is filled with material continuous with the external basal lamina or lamina densa of the muscle cell. This homogeneous material also fills the clefts formed by the infolding of the muscle surface, termed the subsynaptic gutter. This specialisation of the muscle cell membrane presents an increased surface area at this important interface.

The myofibril component of the muscle cell is not seen in this plate, but one muscle cell nucleus is present in its typical position close to the surface of the cell. Adjacent to this are found many mitochondria and other sarcoplasmic components.

CT	Connective tissue space
M	Mitochondrion
N	Nucleus of muscle cell
Ne	Nerve cell
SG	Subsynaptic gutter
V	Vesicles in nerve terminals
X-X	Indicates the point at which the nerve enters into close contact with muscle.

TISSUE Rat skeletal muscle. Osmium fixation, PTA staining.

MAGNIFICATION 21,000 ×.

REFER TO Plates 26, 34a.
Pages 12, 61, 69.

K

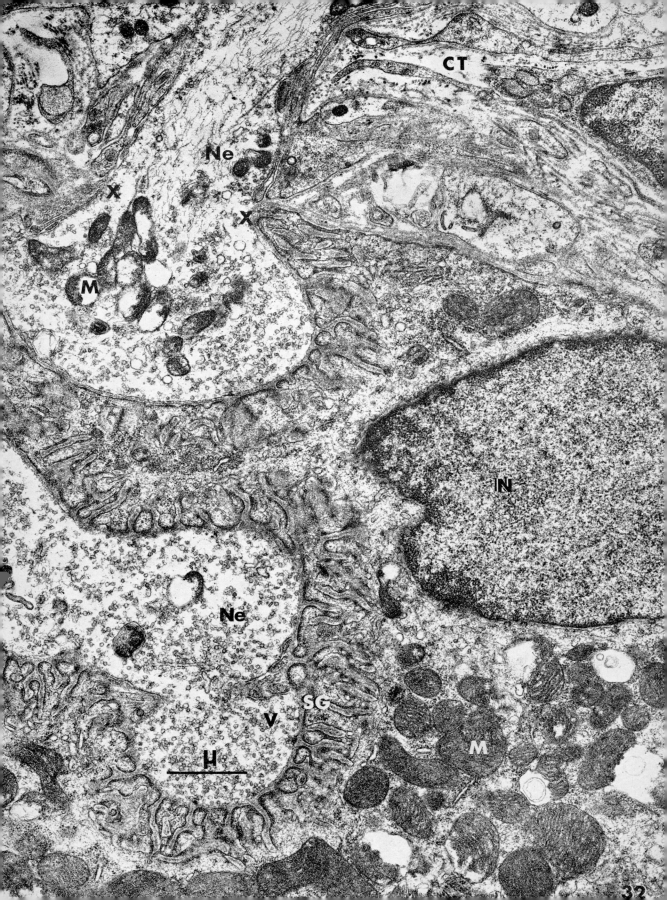

PLATE 33. CEREBELLUM: PURKINJE CELL

This micrograph, at moderate magnification, shows an area of cytoplasm from a Purkinje cell of the cerebellum. The cytoplasm of a nerve cell such as this is not characterised by any specific fine structural appearances, but contains extensive cytoplasmic membrane systems suggesting a high metabolic turnover. The granular endoplasmic reticulum is particularly prominent, the dilated cisternae perhaps reflecting slight fixation damage. There are numerous ribosomes in the ground substance of the cytoplasm leading to the production of areas of marked electron density. Corresponding patches showing cytoplasmic basophilia on light microscopy are known as Nissl bodies. The Golgi apparatus is also prominent and shows the usual features. The mitochondria are of moderate size and are relatively plentiful. There is a lysosome-like structure and a vesicle-containing body. Microtubules are present in the cytoplasmic ground substance. The appearances suggest a significant protein synthetic or secretory activity, but in what way this relates to neuronal function is not clear.

G	Golgi apparatus
GER	Dilated cisterna of granular endoplasmic reticulum
L	Lysosome-like body
M	Mitochondrion
R	Ribosomes
T	Microtubules
VCB	Vesicle-containing body

TISSUE Rat cerebellum. Glutaraldehyde fixation, uranium staining.

MAGNIFICATION 29,000 ×.

REFER TO Plates 7, 9a, 9b, 10, 38.
             Pages 22, 33, 68.

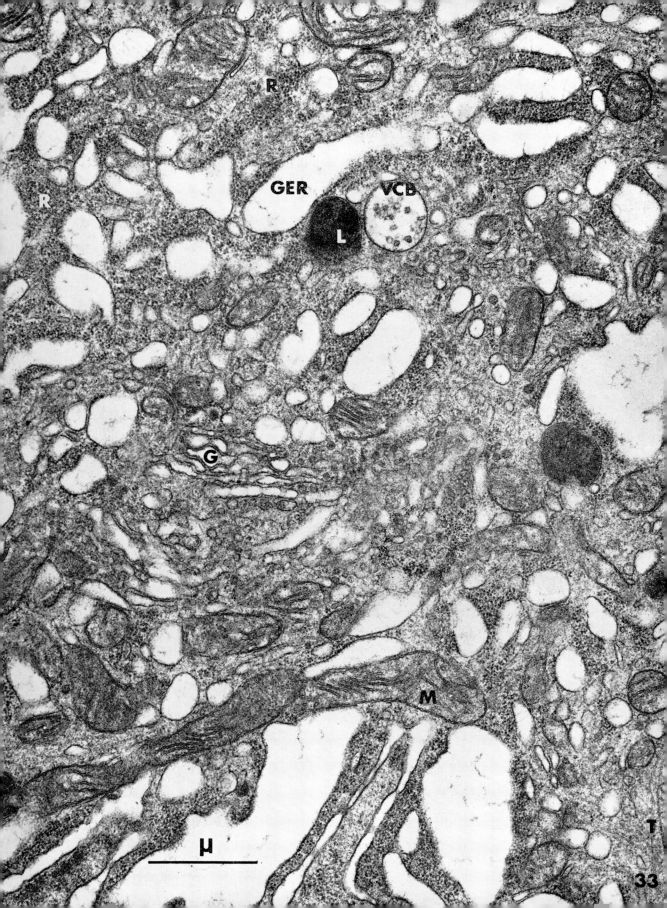

PLATE 34a. PERIPHERAL NERVE: MYELINATED AND UNMYELINATED AXONS

This low magnification micrograph shows a transverse section of part of a mixed peripheral nerve. There are parts of three myelinated nerve axons at the edges of the micrograph as well as a bundle of unmyelinated axons in the centre of the field.

The myelinated nerve axons are surrounded by the dense fatty lamellated myelin sheath which appears to be formed from the surface membrane of the Schwann cell. In the myelinated nerve a single Schwann cell sheath is associated with a single axon. The fully formed myelin layer displaces the remaining Schwann cell cytoplasm to the periphery where it forms a thin rim around the myelinated nerve as indicated on the left of the plate. The axon within the myelin sheath shows neurotubules and neurofibrils in transverse section.

A number of unmyelinated axons are carried within a single Schwann cell sheath. The individual axons are separate from each other, each one being carried within its own groove or tunnel in the Schwann cell, surrounded by the invaginated Schwann cell surface membrane. The mesaxon which suspends each axon within the Schwann cell can be seen at several points communicating with the external surface of the Schwann cell. The cytoplasm of the Schwann cell is dense in appearance with many ribosomes.

Surrounding each Schwann cell is an external basal lamina or lamina densa, indicated at several points by arrows. This is a layer similar to that which surrounds muscle and capillary endothelium and underlies epithelia of all kinds. The basal lamina forms a partition between the cell and the surrounding connective tissue elements. The small dense circular profiles between the cells are cross sections of collagen fibres.

A	Axon
C	Collagen fibres in transverse section
S	Schwann cell cytoplasm
T	Neurotubules in transverse section
X	Mesaxon at Schwann cell surface
Arrows	Indicate the basal lamina surounding each Schwann cell.

TISSUE Rat peripheral nerve. Glutaraldehyde fixation, uranium staining.

MAGNIFICATION 32,000 $\times$.

REFER TO Plates 35, 36, 38.
Pages 12, 56, 70.

PLATE 34b. PERIPHERAL NERVE: MYELIN

This is a high magnification micrograph of myelin. The laminated pattern is clearly shown. The major dense lines are 120 Å apart, separated by an intermediate less dense spacing. The minor dense line is produced by the fusion of the outer laminae of the trilaminar structure of the Schwann cell membrane, the major dense line from the fused inner laminae. The pattern appears to form during the process of myelination by the close apposition of successive layers of the mesaxon around the axon.

TISSUE Rat peripheral nerve. Glutaraldehyde fixation, uranium staining.

MAGNIFICATION 120,000 $\times$.

REFER TO Plate 34a.
Page 71.

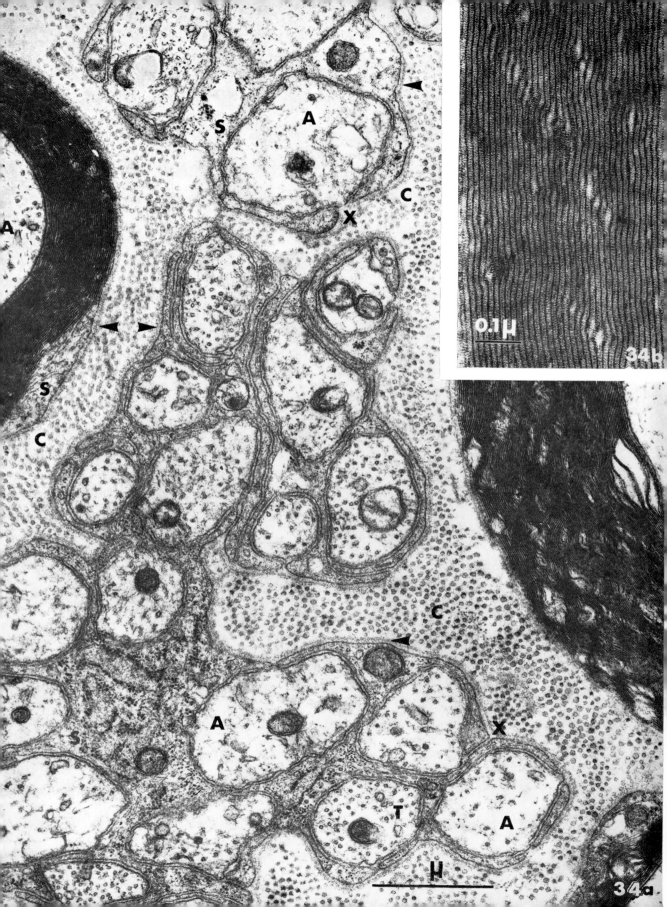

0.1μ

34b

μ

34a

PLATE 35. UNMYELINATED NERVE

This is a micrograph at moderate magnification of unmyelinated nerve cut in cross section. The axons, surrounded by Schwann cell cytoplasm, are suspended by mesaxons formed from the surface membrane of the Schwann cell. Points of communication between mesaxons and the cell surface are indicated. The Schwann cell is surrounded by the basal lamina or lamina densa which separates it from the connective tissue.

Some of these unmyelinated nerve axons come close to the surface of the Schwann cell, coming into contact with the basal lamina at points marked X. One axon, distinguished by the presence of small vesicles, occupies a shallow groove in the Schwann cell rather than a tunnel, suggesting that it is close to its termination. Within the axons are seen cross sections of neurotubules, which appear as small circular profiles, as well as neurofibrils and mitochondria. The surrounding connective tissue is poorly organised with few collagen fibres.

A	Axon
BL	Basal lamina
CT	Connective tissue space
M	Mitochondrion
Mes	Mesaxon
SC	Schwann cell cytoplasm
T	Neurotubules in transverse section
V	Vesicles
X	Axon in contact with basal lamina
Arrow	Indicates a collagen fibre.

TISSUE Rat heart. Glutaraldehyde fixation, uranium staining.

MAGNIFICATION 39,000 $\times$.

REFER TO Plates 21a, 34a, 37.
Pages 12, 33, 56, 70.

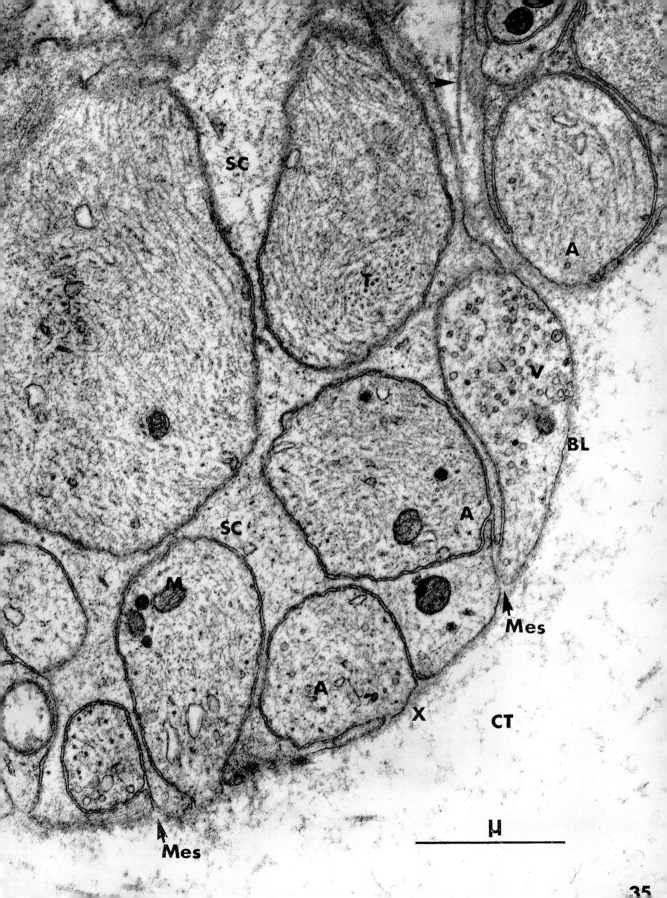

35

PLATE 36. CEREBELLUM: CAPILLARY

This plate shows, at low magnification, a capillary within the cerebellum surrounded by processes of nerve and glial cells. The lumen of the capillary contains a flocculent precipitate of the plasma proteins surrounding part of a red blood corpuscle with its typical dense homogeneous appearance. The endothelial lining of the capillary is of moderate thickness and contains no fenestrations. The elongated endothelial nucleus is present. Two arrows point to the close-fitting basal lamina or lamina densa which surrounds the capillary.

Although at first sight the capillary appears to be surrounded by a clear space which separates it from the nervous tissues, closer inspection shows that these are pale cytoplasmic processes of neuroglial cells which form a sheath around the capillary. The presence of occasional cytoplasmic organelles and fine fibrillar material indicates the cellular nature of this pale sheath, composed of several different foot processes. The contacting glial cell membranes which are seen at CM are an indication of the close packing of these processes. This sheath effectively isolates the capillary endothelium and its basal lamina from the surrounding nervous tissues and may be a significant factor in the maintenance of the blood-brain barrier.

The surrounding nervous tissue is very complex. Many of the individual tiny processes which are seen in the electron micrograph are far below the limit of resolution of the light microscope. A number of synaptic structures can be distinguished at different points in this field. Notice the complete absence of a true connective tissue space in brain tissue. Collagen fibres are not found between nerve cells. The neuroglial cells support the nerve cells and may control metabolic exchanges. There is virtually no extracellular space in the brain on conventional electron microscopic examination, individual processes of nerve and glial cells being in close contact, separated by only 150 to 200 Å.

AC	Process of a neuroglial cell, probably an astrocyte
CM	Cell membranes of adjacent neuroglial foot processes
M	Mitochondrion
N	Capillary endothelial nucleus
RBC	Red blood corpuscle
Arrows	Indicate the basal lamina or lamina densa of the capillary sandwiched between the endothelial cell and the neuroglial foot processes.

TISSUE Rat cerebellum. Glutaraldehyde fixation, PTA staining.

MAGNIFICATION 12,000×.

REFER TO Plates 17a, 17b, 17c, 33, 34a, 37, 38.
Pages 49, 54, 72.

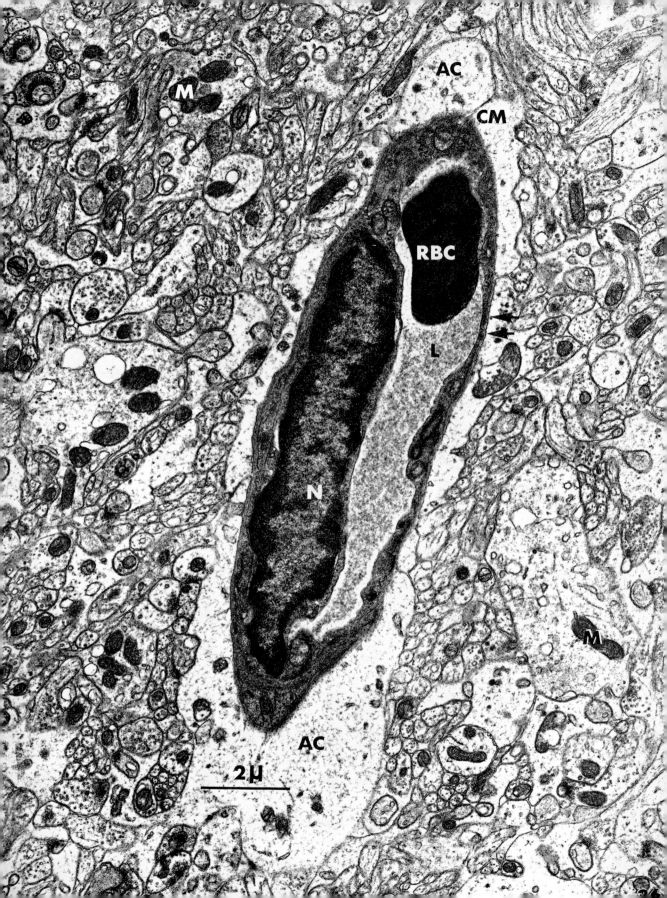

PLATE 37. CEREBELLUM

This micrograph is of an area of the cerebellum showing the complexity, revealed at moderate magnification, of the central nervous system. The close packing of cell processes and the absence of intercellular spaces of significant dimensions can be shown clearly only by the electron microscope. Bundles of minute nerve processes, each a fraction of a micron in diameter, are closely packed between neuroglial cells and synaptic terminals. Adjacent processes are separated by a pale interspace of about 150 Å in width. This apparent space may, of course, be occupied by some intercellular material, perhaps of mucopolysaccharide nature, which cannot be distinguished with the present techniques.

Several typical synapses are seen in this micrograph. The presynaptic terminals contain synaptic vesicles and membrane thickenings are found in both the pre- and postsynaptic terminals. The presence of a gap at the synapse which must be crossed by transmitter substance accounts for some of the physiological properties of the synaptic junction. At the lower left hand corner of the micrograph the synapse indicated by SV appears to be double, two postsynaptic terminals being in contact with one set of presynaptic vesicles.

AC	Cytoplasm of neuroglial cell, probably an astrocyte
M	Mitochondrion
SV	Synaptic vesicles
X	Small nerve process lying in a bundle of processes of similar size

TISSUE Rat cerebellum. Glutaraldehyde fixation, PTA staining.

MAGNIFICATION 34,000 ×.

REFER TO Plates 32, 36.
Pages 10, 69, 72.

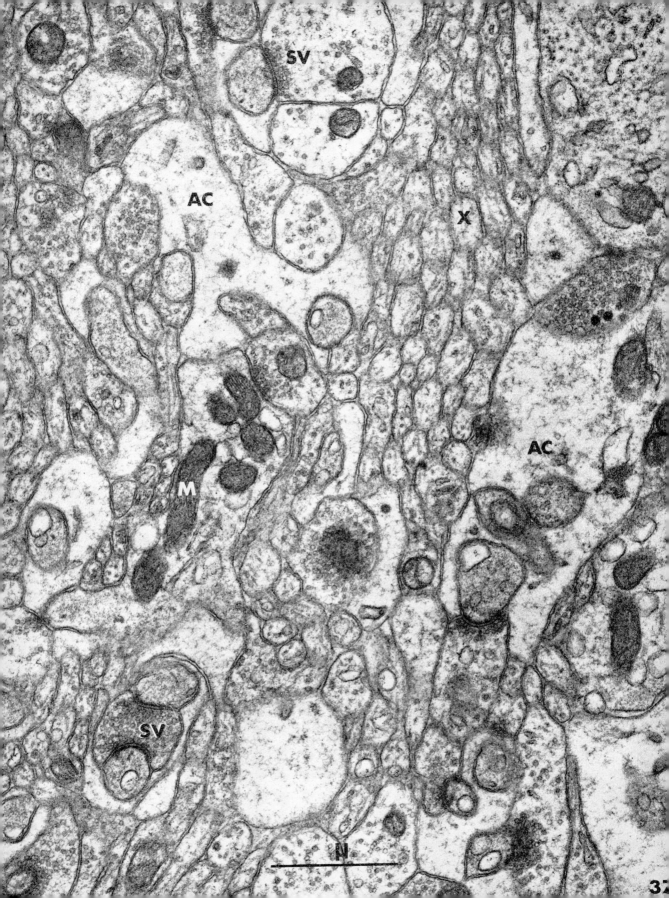

PLATE 38. NEUROTUBULES: PURKINJE CELL

This is a micrograph at moderate magnification of a dendrite of a Purkinje cell in the cerebellum. The Purkinje cell process crosses the field vertically, occupying most of the plate, but is surrounded by smaller closely packed cell processes.

Within the Purkinje cell there are numerous neurotubules, lying parallel to the long axis of the process. These structures are the same as the microtubules seen in many other cells. The individual tubules extend over considerable distances apparently unbroken and unbranched. Components of this type are probably partly responsible for producing the images of neurofibrils as shown by silver staining methods for light microscopy. Mitochondria and smooth surfaced vacuoles are also seen.

Between the Purkinje cell and adjacent cell processes there is virtually no extracellular space, other than the narrow cleft always seen between contacting cells. The small processes of nerve cells, characterised by the presence of synaptic vesicles in some cases, are often surrounded by processes of neuroglial cells, their pale cytoplasm suggesting that they are astrocytes. The neuroglial cytoplasm surrounding the nerve cells may act as a channel for diffusion through the brain tissues, to and from the capillaries which have close neuroglial connections. Such metabolic exchange would therefore be under cellular control instead of being dependent simply on the physical constants of a connective tissue space.

AC   Neuroglial cell, probably astrocyte
M   Mitochondrion
SV   Synaptic vesicles
T   Neurotubules
V   Vacuole of smooth endoplasmic reticulum

TISSUE Rat cerebellum. Glutaraldehyde fixation, uranium staining.

MAGNIFICATION 39,000 ×.

REFER TO Plates 33, 34a, 36, 43.
         Pages 34, 68.

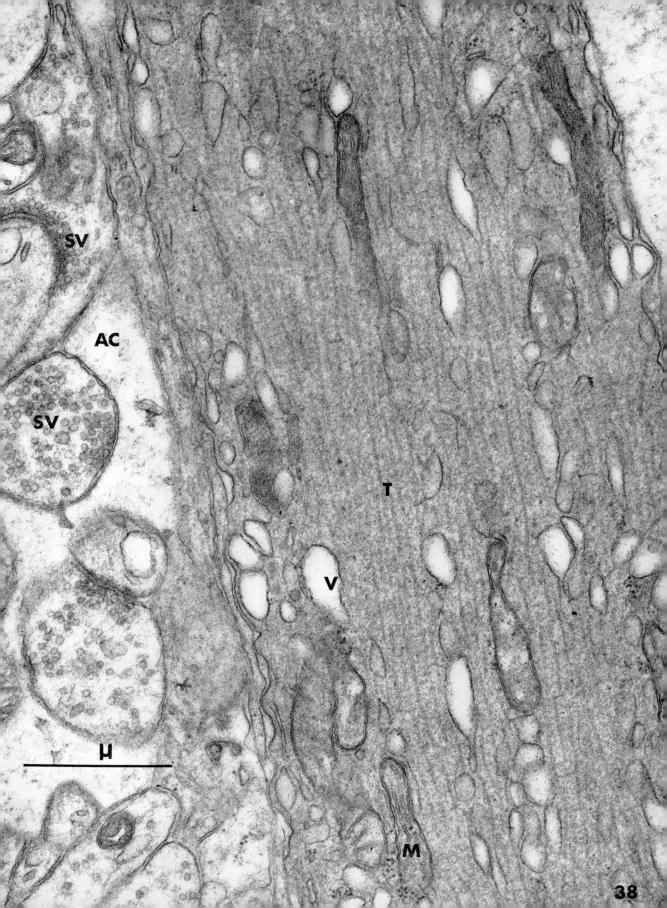

38

PLATE 39. PERMEABILITY: RENAL GLOMERULUS

This low magnification micrograph shows part of Bowman's capsule with glomerular capillaries, and part of an adjacent renal tubule. The glomerular capillaries show the typical thin endothelial walls in which pores or fenestrations can be seen. The related basal lamina separates the endothelium from the surrounding foot processes of the podocytes. The cell bodies of several podocytes lie between the capillaries. The closely packed rows of minor foot process are easily seen at Y, where there is a direct passage from the capillary between the foot processes to the urinary space of Bowman's capsule. At X, however, an apparent sub-podocytic space is enclosed by part of the podocyte cytoplasm. The sub-podocytic space is not in direct continuity with the urinary space in this section, although it may open into it in another plane. The podocytes are the visceral epithelial cells of Bowman's capsule.

On the left of the field the parietal epithelium of Bowman's capsule is seen. This is a thin layer of cells which becomes continuous with the podocytes at a point not demonstrated in this plate. This layer of cells lies on a thick basal lamina. A connective tissue space separates Bowman's capsule from the nearby renal tubule. The base of the tubular cells can be seen to have marked basal infolding indicating membrane-associated metabolic activity, probably related to active transport mechanisms.

BL	Basal lamina
CT	Connective tissue space
En	Endothelium
Ep	Visceral epithelium of Bowman's capsule; the podocytes
F	Basal infoldings of kidney tubule cell
L	Lumen of capillary
N	Nucleus of endothelial cell
P	Endothelial pores
PEp	Parietal epithelium of Bowman's capsule
US	Urinary space of Bowman's capsule
X	Apparent sub-podocytic space
Y	Area of direct communication between glomerular capillary and urinary space

TISSUE Mouse kidney. Osmium fixation, uranium staining.

MAGNIFICATION 20,000 ×.

REFER TO Plates 17c, 18, 19, 20, 40, 41a.
　　　　　Pages 7, 12, 47, 49, 52.

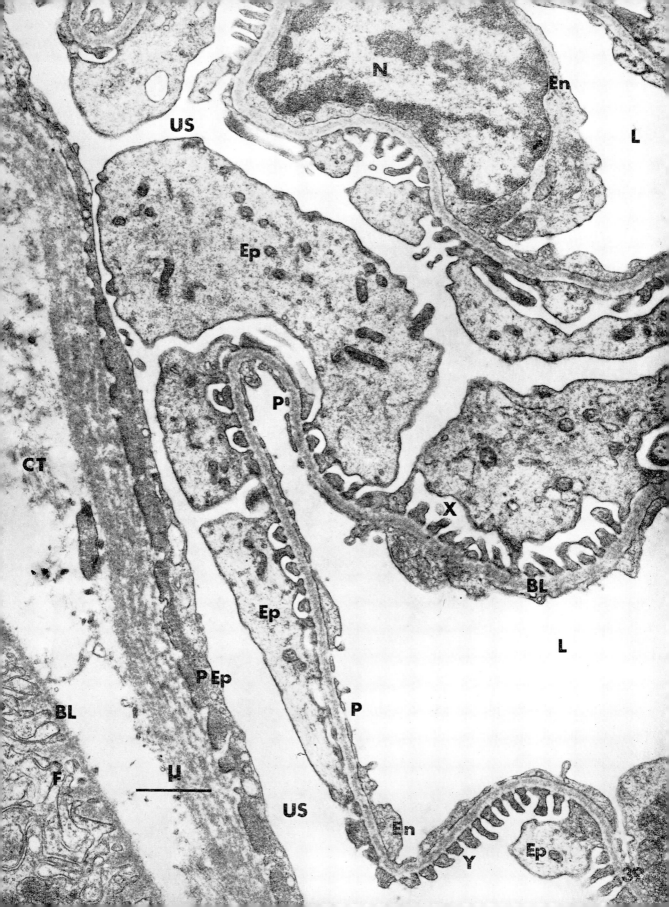

PLATE 40. PERMEABILITY: RENAL GLOMERULUS

This is a high magnification micrograph showing the barrier between the lumen of the glomerular capillary and the urinary space of Bowman's capsule. This plate shows a small part of Plate 39 at a higher magnification.

The endothelial cell which lines the capillary has pores or fenestrations in its thin wall. The basal lamina is a moderately dense, almost homogeneous layer, in which the only detail seen is a suggestion of a fibrillar or filamentous network. The basal lamina separates the endothelial and the epithelial components of the renal glomerulus. Each podocyte foot process, or pedicel, is covered by the podocyte surface membrane which has the typical trilaminar membrane structure shown elsewhere. The filtration slit membrane, indicated by an arrow, the final barrier between blood and urine, links adjacent foot processes. It does not have a trilaminar substructure.

When the glomerulus is the site of disease, the foot processes may become fused and thickened and the basal lamina becomes often irregular in outline. Such changes may be reversible on treatment.

BL   Basal lamina
Cap   Lumen of glomerular capillary
En   Endothelium
Ep   Glomerular epithelial cell; prodocyte
FP   Minor podocyte foot processes or pedicels
P   Endothelial pore
US   Urinary space
Arrow   Indicates filtration slit membrane between adjacent podocyte foot processes.

TISSUE Mouse kidney. Glutaraldehyde fixation, uranium staining.

MAGNIFICATION 76,000 ×.

REFER TO Plates 18, 20, 39.
      Pages 5, 12, 49, 52.

L

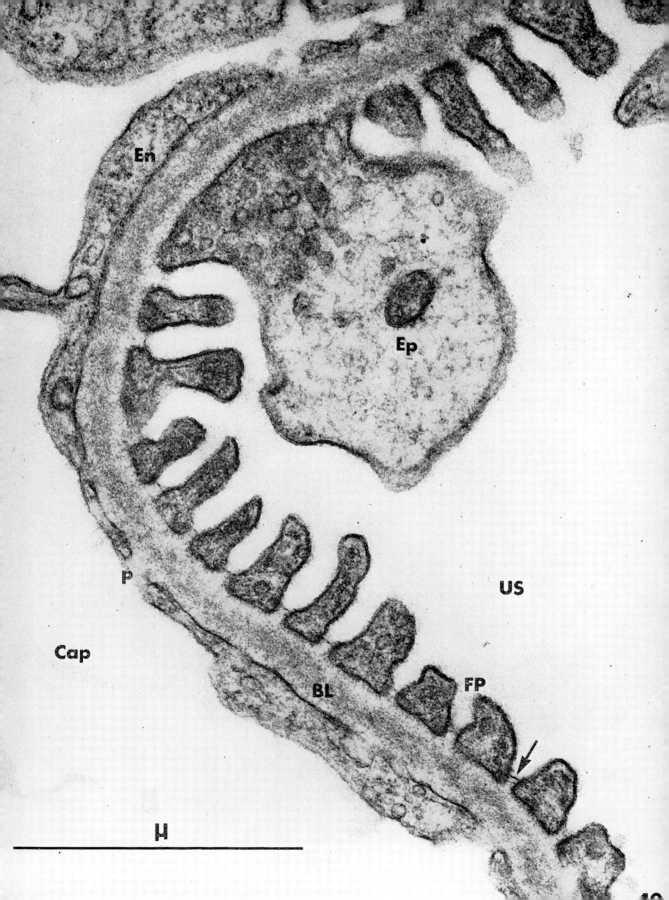

PLATE 41a. PERMEABILITY: RENAL GLOMERULUS

This is a high magnification micrograph of the basal lamina and related structures in the renal glomerulus. The endothelium of the glomerular capillary is very thin and is perhaps cut obliquely in places. The endothelial fenestrations are not clearly seen. The basal lamina appears homogeneous.

The podocyte foot processes lie close together in contact with the outer side of the basal lamina. The individual minor foot processes, indicated by P, are not completely separate, being joined by a tenuous filtration slit membrane. The composition of this barrier is not known, but it presents a significant block to the passage of certain materials. One major foot process crosses the field, enclosing an apparent sub-podocytic space, in which the arrow lies, between itself and the row of minor foot processes. The extent to which this space communicates freely with the main urinary space of Bowman's capsule cannot readily be determined from a single electron micrograph. The cytoplasm of the podocyte is of moderate density. A feature of these cells on close inspection is the presence of many microtubules commonly associated with an asymmetric cell shape. The microtubules are better preserved with glutaraldehyde fixation than with osmium.

BL	Basal lamina
En	Endothelium
Ep	Glomerular epithelial cell; podocyte
L	Lumen of capillary
P	Podocyte minor foot process
RBC	Red blood corpuscle
T	Microtubules
Arrow	Indicates filtration slit membrane, linking adjacent foot processes.

TISSUE Mouse kidney. Osmium fixation, uranium staining.

MAGNIFICATION 33,000 ×.

REFER TO Plates 18, 20, 38, 39, 40.
          Pages 5, 12, 49, 52.

PLATE 41b. PHOTORECEPTOR: CHLOROPLAST

This high magnification micrograph shows part of a chloroplast in which several groups of membrane lamellae are set in a fine granular background. This plate does not show the highly organised pattern of membrane packets or grana, seen in the chloroplasts of some green plants. It does however illustrate the essential feature of the active plant photoreceptor, the presence of membrane lamellae which trap light energy and convert it into chemical energy through the action of associated photosensitive molecules. The essential pigment component of the chloroplast is chlorophyll, without which the photosynthetic reactions, combining carbon dioxide and water to produce carbohydrate, cannot proceed.

TISSUE Lettuce leaf. Osmium fixation, uranium staining.

MAGNIFICATION 150,000 ×.

REFER TO Plates 42a, 42b.
          Pages 73, 74.

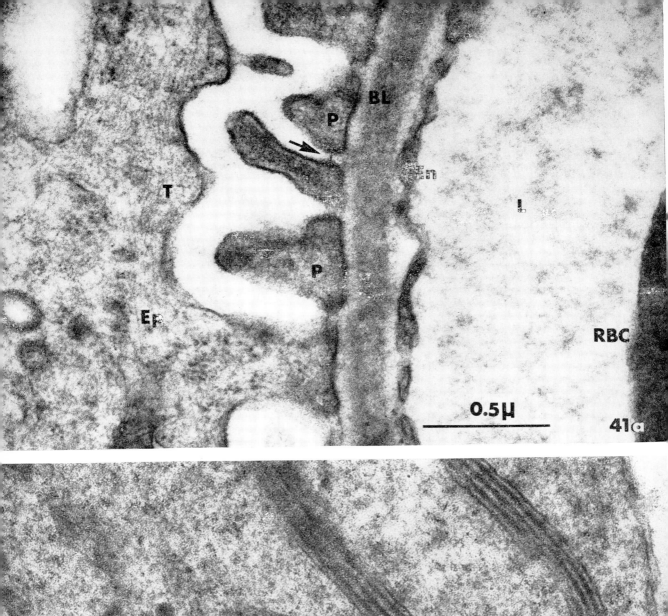

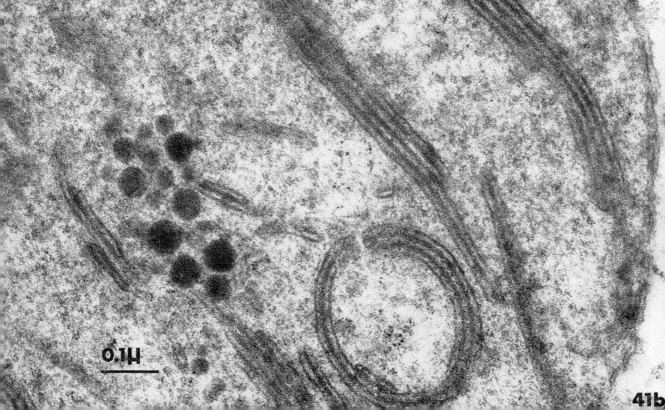

PLATE 42a. PHOTORECEPTOR: RETINA

This is a high magnification micrograph of the outer segment of a retinal rod photoreceptor. The cell membrane, which forms the limit to the structure, is seen at the left of the field. The closely packed lamellae which fill this part of the cell are the location of the photo-sensitive pigment-protein mechanism which converts the energy of light into cell activity. Maximum light-trapping potential is attained by the packing of the photoreceptor structure with the active membrane system concerned with its specific function, to the exclusion of other components.

TISSUE Rat retina. Osmium fixation, lead staining.

MAGNIFICATION 146,000 ×.

REFER TO Plates, 41b, 42b.
        Pages 6, 73, 74.

PLATE 42b. PHOTORECEPTOR: RETINA

This plate shows the junction between the retinal rod outer and inner segments at low magnification. The closely packed membrane lamellae of the outer segment can be seen. The inner segment, in contrast, contains the normal cytoplasmic components including ribosomes and mitochondria. There is a pair of centrioles present in the inner segment, one of which gives rise to a structure with the morphology of a cilium. This short cilium connects the outer segment with the inner segment. It may serve the function of communication between the two segments.

TISSUE Rat retina. Osmium fixation, lead staining.

MAGNIFICATION 30,000 ×.

REFER TO Plates 12a, 12b, 13b, 43, 44a, 44b.
        Pages 6, 35, 57, 73, 74.

Ce	Centrioles
CM	Cell membrane surrounding the retinal rod
IS	Inner segment
OS	Outer segment
R	Ribosomes

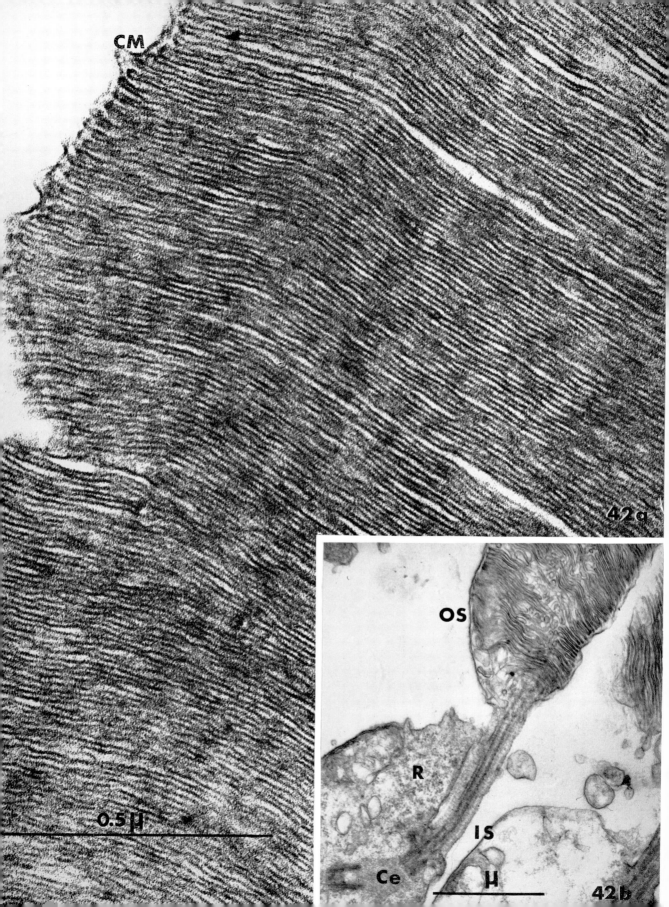

CM

0.5 μ

OS

R

IS

μ

Ce

42a

42b

PLATE 43. MOVEMENT: CILIA

This is a micrograph at moderate magnification of the surface of a tracheal epithelial cell showing the cilia with their internal axial filament complexes ending in basal bodies. Short rootlets anchor the cilia in the apical cytoplasm. In the upper part of the micrograph the cilia have been cut obliquely and at times nearly transversely, showing to some extent their typical 9+2 pattern of organisation. In the lower parts of the cilia the axial components are sectioned longitudinally, showing their tubular nature. Notice the dimensions of the cilia, compared with the size of the few microvilli which are also seen at the cell surface.

The surface membrane of the tracheal epithelial cell covers the entire cilium as well as the adjacent microvilli. A junctional complex marks the contact point between adjacent cells at the left of the field, but obliquity of section obscures its morphology. The arrows in the cytoplasm indicate points along a microtubule which lies in the plane of the section. Its diameter is comparable with that of the axial components of the cilia.

B	Basal body
C	Cilia
JC	Junctional complex
L	Lumen of trachea
MV	Microvillus
R	Rootlet of cilium
Arrows	Indicate a microtubule in the apex of the tracheal epithelial cell.

TISSUE Rat trachea. Glutaraldehyde fixation, uranium staining.

MAGNIFICATION 45,000 ×.

REFER TO Plates 4a, 5, 44a, 44b, 45, 46.
Pages 33, 34, 57.

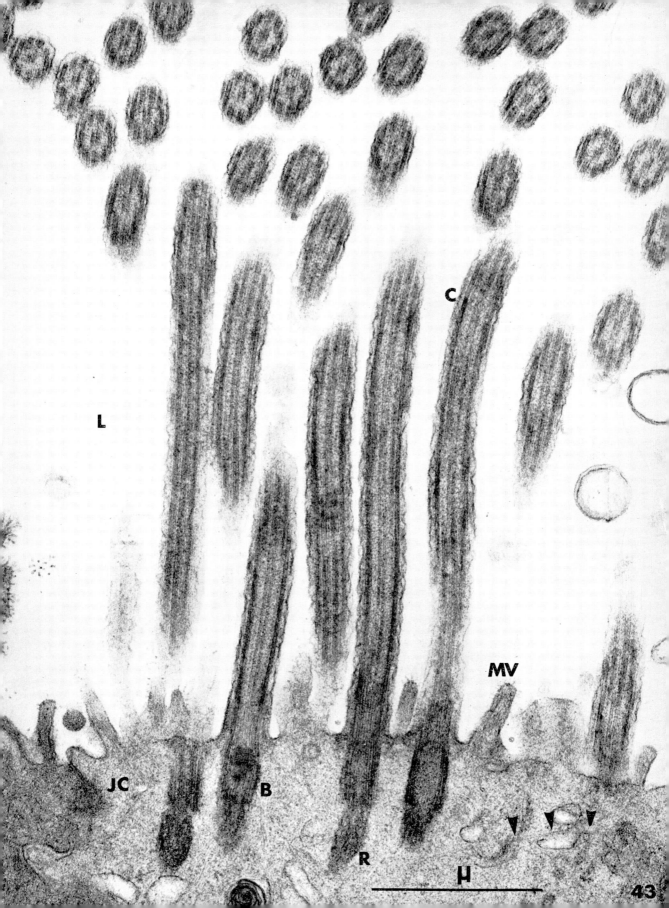

43

PLATE 44a. MOVEMENT: CILIA

This is a high magnification micrograph of part of a single cilium cut in longitudinal section. The cilium is covered by the surface membrane of the apex of the tracheal epithelial cell, its trilaminar structure being visible at different points. The same membrane covers the adjacent small microvilli. The arrow indicates the radiating filaments which constitute the cell coat or fuzzy coat, the appearance presented here being sometimes described as antennulae microvillares. These projections, attached to the surface of the outer lamina of the cell membrane, are thought to be of mucopolysaccharide nature.

The differences between cilia and microvilli are clearly shown in this plate. The cilium has a complex central core which is not seen in the microvillus. The longitudinal elements of the core, the so-called axial filament complex, which terminate in the basal body, are tubular in construction. The two central tubules of the cilium lie in the plane of section in the upper part of the cilium shown here. Each cilium is thick enough to be individually visible by light microscopy. A single microvillus is too small to be resolved. Cilia are motile, microvilli non-motile.

AFC   Axial filament complex
B   Basal body
C   Cilia
L   Lumen of trachea
MV   Microvillus
Arrow   Indicates 'antennulae microvillares', projecting from the surface of the microvillus.

TISSUE Rat trachea. Glutaraldehyde fixation, uranium staining.

MAGNIFICATION 130,000 ×.

REFER TO Plates 5, 25a, 43, 45, 46.
        Pages 5, 12, 36, 57.

PLATE 44b. MOVEMENT: CILIA

This micrograph shows a transverse section at moderate magnification through the cilia and microvilli at the apex of a tracheal epithelial cell. The internal structure of the cilia can now be seen. The nine peripheral pairs of tubules surrounding the central pair can all be made out. The roughly parallel arrangement of the central pair in adjacent cilia is indicated by lines. The plane of ciliary beat is at right angles to this alignment. The differences between cilia and microvilli are made clear in this micrograph.

Lines   Indicate direction of alignment of the central pairs of tubules in the axial filament complex of related cilia.

TISSUE Rat trachea. Glutaraldehyde fixation, uranium staining.

MAGNIFICATION 51,000 ×.

REFER TO Plates 5, 12b, 43, 45.
        Pages 36, 57.

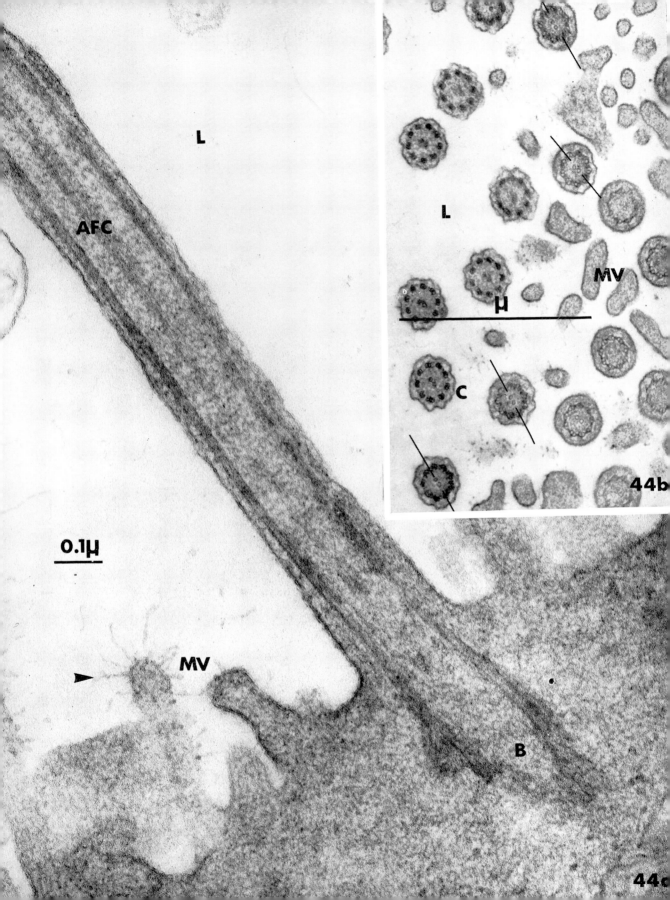

L

AFC

0.1μ

MV

B

44a

44b

L

MV

μ

C

MV

PLATE 45. MOVEMENT: SPERMATOZOON

This micrograph shows a sperm tail cut in transverse section, at high magnification. The various components of the tail are clearly seen. The cell membrane of the spermatozoon forms a continuous surrounding layer as in any other cell. The centre of the sperm tail contains the typical 9+2 arrangement of subunits, similar to that seen on cross section of cilia. The components of this axial filament complex are tubular in nature.

In the centre of the tail the two central tubules are clearly seen, separated by a narrow space. The surrounding nine pairs of tubules are evenly spaced around the central pair. The two components of each pair of peripheral tubules lie close together. One of these, subfibril A, has a denser central core, the other, subfibril B, has a pale central core. Two short, diffuse arms extend from subfibril A to subfibril B of the next doublet. The arms of successive doublets point in this case in a clockwise direction, indicating that the sperm is being viewed, in this cross-section, from head to tail.

Peripheral to the nine doublets of the axial filament complex are the nine large dense additional components usually present in mammalian spermatozoa. They have an asymmetrical arrangement and can be numbered definitively, the solitary large component being designated component 1, the others being numbered in sequence from this point in the direction indicated by the arms extending from subfibril A of the axial filament complex. Two other large components are consistently seen at points 5 and 6. Components 3 and 8, situated close to the axis of the tail as defined by the line joining the centres of the two central filaments, are small. They do not extend as far towards the tip of the tail as the other components. These dense elements may have a contractile function, reinforcing the action of the peripheral doublets of the axial filament complex to which they are related.

There is a dense 'fibrous' sheath surrounding the components described above, replacing the mitochondrial spiral which is found in the mid-piece of the sperm. This sheath is composed of 'ribs' which are reinforced in the axis of the sperm by thickenings which represent a longitudinal 'backbone'. The full significance of the various components of the sperm tail is not yet clearly known.

A	Subfibril A
B	Subfibril B
CM	Cell membrane
DB	Dense additional component of sperm tail
FS	Fibrous sheath, with thickened portion
1, 2, 3	Indicate the first three additional components numbered in sequence in the direction pointed by the arms projecting from subfibril A.

TISSUE Spermatozoon from rat epididymis. Glutaraldehyde fixation, uranium staining.

MAGNIFICATION 220,000 ×.

REFER TO Plates 38, 43, 44a, 46.
        Page 59.

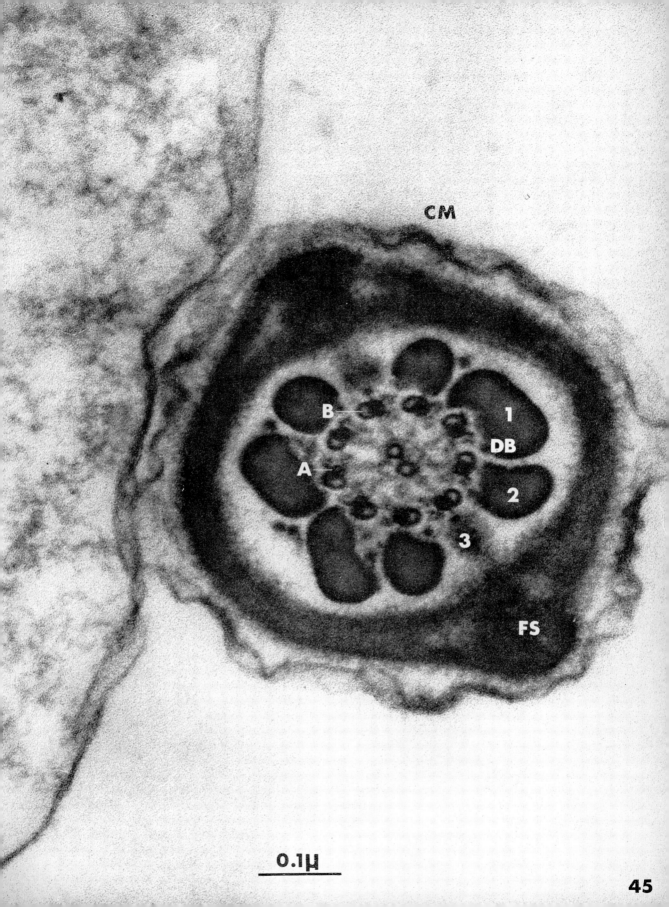

CM

B

A

1

DB

2

3

FS

0.1µ

45

PLATE 46. MOVEMENT: SPERMATOZOON

This is a high magnification micrograph of a sperm tail cut in longitudinal section. The limiting membrane of the sperm shows at points a trilaminar construction. The 'fibrous' sheath lies within the limiting membrane. The central axial filament complex consists of microtubular components similar to those seen in the cilium. The dense additional components peripheral to the axial filament complex are seen in longitudinal section. They display no evident substructure. A fine periodicity is, however, seen along the central pair of the axial filament complex.

AFC   Axial filament complex
DB   Dense additional component of sperm tail
CM   Cell membrane
FS   Fibrous sheath
Arrows   Indicate points where trilaminar construction can be seen in the cell membrane.

TISSUE Spermatozoon from rat epididymis. Glutaraldehyde fixation, uranium staining.

MAGNIFICATION 118,000 ×.

REFER TO Plates 43, 44a, 45.
          Pages 5, 59.

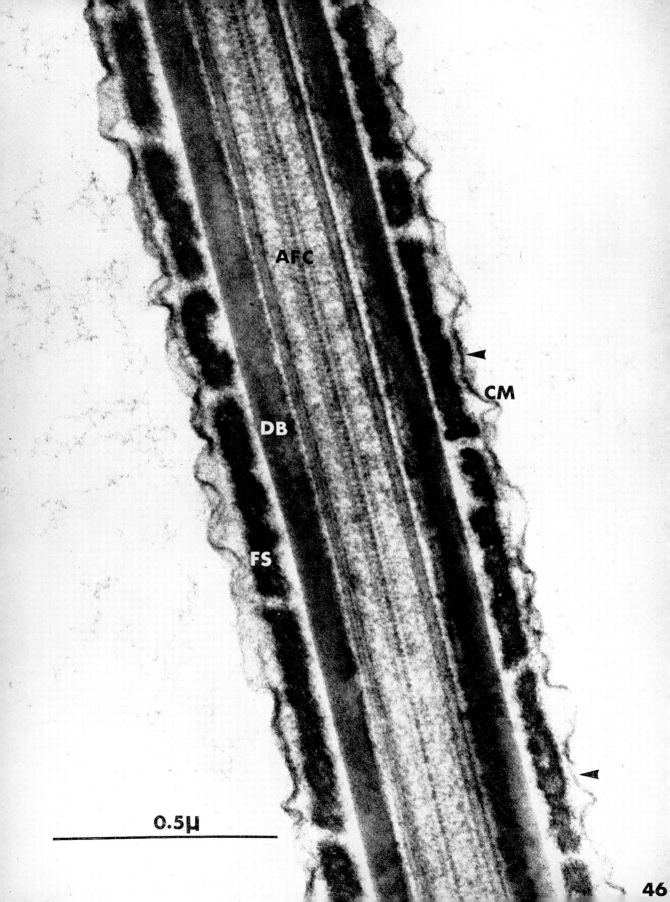

AFC

CM

DB

FS

0.5µ

46

PLATE 47. PLANT CELL WALL

This micrograph, at high magnification, shows the cell wall and adjacent cell membrane of a plant cell. Plants have rigid external cellulose cell walls which lie immediately outside the true cell membrane. The cell walls which give the plant its structural rigidity can be regarded as analogous to the glycocalyx or mucopolysaccharide cell coat. Some form of cell coat is believed to exist in all cells, but is not necessarily visible by microscopic techniques. The cell membrane marks the boundary of the cell and shows the trilaminar construction already demonstrated in membranes of various types in animal cells.

Within the cell are numbers of particles of ribosomal dimensions and appearance. Part of a chloroplast is present at the upper right hand corner of the plate. A periodic structure is seen at X. This appears to be some type of intracellular crystal. Although specialised in various ways, the plant cell shares many fine structural features with the animal cell.

CM   Cell membrane with trilaminar structure
CW   Cellulose cell wall
R   Ribosome-like particles
X   Crystalline component

TISSUE Lettuce leaf. Osmium fixation, uranium staining.

MAGNIFICATION 160,000 ×.

REFER TO Plates 2, 5, 25a, 25b, 41b.
        Pages 5, 11.

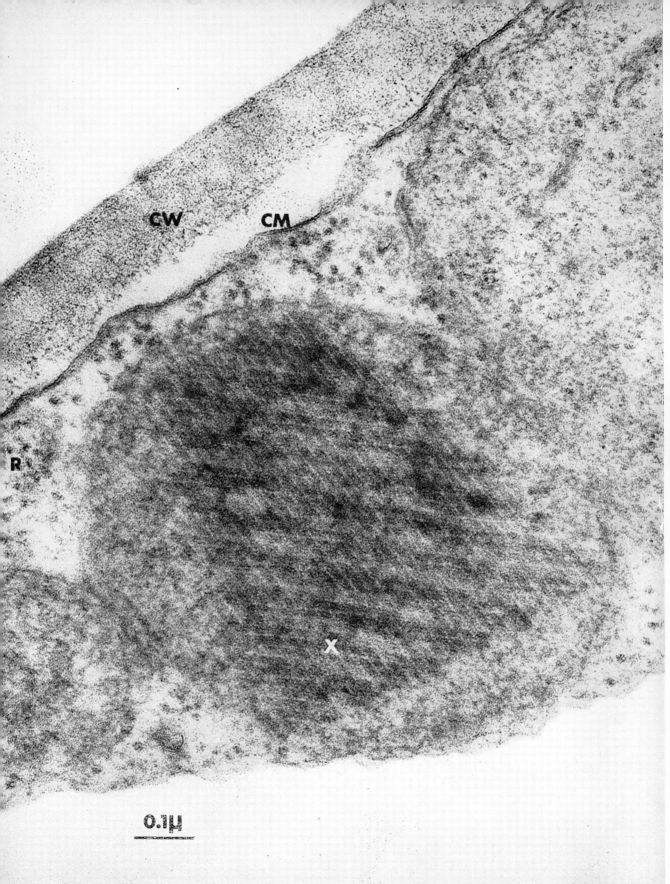

CW

CM

R

X

0.1μ

47

Index

Arabic numerals refer to the text, figures in **bold** type refer to plates

Printed in Great Britain by Neill & Co. Ltd., Edinburgh